Collins Guide to
# Tropical Plants

Collins Guide to

# Tropical Plants

*A descriptive guide to 323 ornamental and economic plants with 274 colour photographs*

Wilhelm Lötschert
Gerhard Beese

*translated by* Clive King

**Collins**
Grafton Street · London

William Collins Sons & Co. Ltd
London · Glasgow · Sydney · Auckland
Toronto · Johannesburg

*Pflanzen der Tropen* was first published in Germany in 1981
by BLV Verlagsgesellschaft mbH, München



ISBN Hardback edition: 0 00 219112 1

Filmset by Servis Filmsetting Ltd, Manchester
Colour reproduction by Kösel GmbH & Co., Kempten
Text printed and bound by William Collins Sons & Co. Ltd, Glasgow

**Picture credits**

| | |
|---|---|
| Apel: | 140, 151 |
| Croockewit: | 199, 201 |
| de Cuveland: | 263, 264 |
| Eisenbeiss: | 132, 133, 152, 155, 157, 159, 160, 161, 162 |
| Eisenreich: | 67, 123, 156, 165, 166, 167, 168, 168a |
| Felten: | 207, 154, 265 |
| Kribben: | 1, 9, 20, 28, 32, 39, 41, 42, 47, 50, 51, 52, 55, 59, 71, 76, 116, 163, 164, 183, 187, 197, 198, 202, 211, 212, 216, 217, 222, 225, 234, 238, 239, 252, 256, 260 |
| Lötschert: | 2, 4, 6, 7, 10, 12, 14, 16, 17, 21, 22, 24, 25, 26, 27, 33, 34, 35, 37, 38, 43, 43a, 44, 46, 53, 56, 57, 58, 60, 61, 64, 65, 66, 68, 69, 70, 72, 73, 74, 75, 78, 79, 80, 82, 83, 84, 88, 90, 91, 92, 94, 97, 98, 99, 100, 103, 104, 105, 106, 114, 115, 120, 121, 122, 124, 125, 126, 127, 128, 129, 130, 134, 135, 136, 138, 139, 144, 146, 150, 153, 170, 171, 172, 173, 174, 175, 177, 178, 179, 180, 182, 184, 186, 188, 189, 190, 192, 193, 196, 203, 205, 206, 208, 209, 210, 221, 223, 224, 226, 227, 228, 229, 231, 235, 237, 240, 241, 243, 247, 255, 257, 267, 268, 269 |
| Rietschel: | 36, 154, 200 |
| Rysy: | 158 |
| Seibold: | 148, 149 |

*All other photographs*: G. Beese

# Contents

# Preface

This book is an introduction to the colourful world of tropical plants. It deals mainly with decorative and economic plants common in all regions of the tropics. Because of their beauty ornamental plants, especially, have been spread by man from their native lands to all parts of the world. We have therefore tried to select the most striking and widespread species from this bewildering profusion.

The book is intended for amateurs as well as for professional botanists, and particularly for those travelling in the tropics who may find themselves in a world of strange plants, not only in gardens and parks, in streets and by the sea, but in markets too. In addition to a botanical description the systematic relationships of each plant are given, together with the meaning of its name, and its flowering time, origin, distribution and ecological requirements. In the case of economic plants details of their use and an account of their distribution and production are also given. The text is headed by the identifying number of the relevant photographs. Using both the pictures and the text it will be possible for the amateur to identify individual species.

The number of tropical species is very large, making it impossible for all species to be included. This is especially true for timber trees and medicinal plants, of which new kinds are still being discovered. This state of affairs makes it clear how important it is to put a stop to further destruction of tropical vegetation. For the sake of clarity ornamental plants have been divided into trees, palms, shrubs, climbers, and herbaceous species. Economic plants have been arranged according to their uses and products. In the naming of plant species the scientific name is essential, for these plants may be known by several common names. In a number of cases difficulties arise with English names since the plants do not occur in Britain. However, as the native name for the same plant may vary from country to country, the English name has been given whenever possible.

In spite of the necessary limitations imposed by the choice of species we hope that the selection is representative, and that the book will be useful to many plant-lovers on their tropical journeys, and will reveal to them the strange and fascinating variety as well as the beauty of exotic plants.

# Introduction

## Ornamental and Economic Plants in the Tropics

The tropical zone is the home of a vast number of ornamental and economic plants, many of them unfamiliar or unknown to Europeans. However, since the great voyages of discovery a considerable number of tropical economic plants have become some of the most sought after natural products of those regions. There are common spices such as pepper, nutmeg, cinnamon and cloves. Then there is a wide range of beverages and spices, including coffee, tea, cocoa, cola, maté-tea, ginger, curry and allspice. Tropical fruits such as bananas and pineapples are on sale nowadays just like our native fruits. Coconuts, Brazil nuts and peanuts are also of tropical origin. Even the potato, which originated in the uplands of the Andes, is a tropical plant, although few people remember that today.

The discovery of tropical economic and decorative plants together with their export and import has a long and varied history. Because of its great economic value such exploitation frequently resulted in wars and diplomatic complications. The introduction into Europe of unknown and at first very valuable ornamental plants, and their cultivation in glasshouses for the first time, were often sensational events, and owning them has been regarded through the centuries as a special privilege.

With the development of modern forms of transport, tropical fruits and ornamental plants are increasingly common freight in our latitudes. Their appearance in supermarkets and flower shops is visible proof of this. Exotic fruits such as the mango and avocado, or even passion-fruit, are nowadays no longer out of the ordinary.

The increasing wave of tourism to distant places has now reached the homes of these species and many visitors return with an impression of the variety that exists in the world of tropical plants. The purpose of this book is to present an overall view of that world.

## Area and Climate of the Tropics

The region of the tropics lies between the Tropic of Cancer and the Tropic of Capricorn, 23° north and 23° south of the equator respectively. The tropical climate is characterised by uniformly high temperatures and high humidity. The average annual temperature at sea level varies only slightly around 26°–27°C, and the extreme values are rarely below 18° or above 35°C. The relative air humidity lies between 70% and 80%, a very high average.

The equatorial region contains the area with the world's greatest annual rainfall which is spread over the whole year. It forms a belt some 10° north

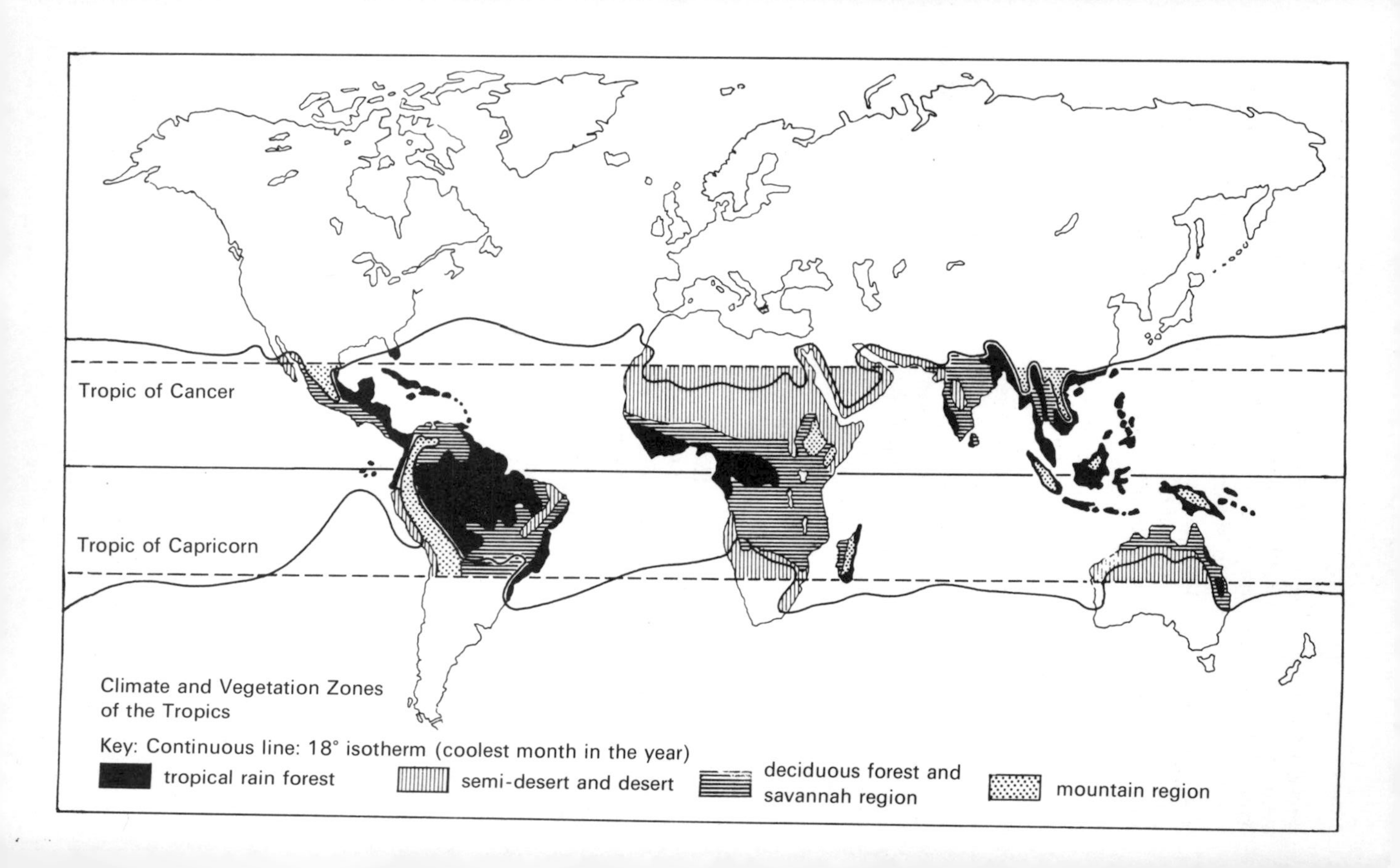
Tropic of Cancer
Tropic of Capricorn
Climate and Vegetation Zones of the Tropics
Key: Continuous line: 18° isotherm (coolest month in the year)
tropical rain forest
semi-desert and desert
deciduous forest and savannah region
mountain region

and south of the equator. The rainfall can be considerable especially on the slopes of high mountains. For example, in the Gulf of Guinea south-west of Mt Cameroun an average annual rainfall of 11,000mm has been recorded.

It has generally been established that rainfall in the tropical zone is greatest at heights of 1000–2000m and then declines again. The temperature also drops with the increase in height, at the rate of about 0.4–0.5°C for every 100m of elevation. Frosts occur occasionally above 1800–2000m and regularly above 3000m.

A feature of the tropical climate, in contrast to that of Europe, is the almost total lack of fluctuation in the temperature throughout the year. Because of the relatively uniform temperature on every day in the year, the tropical climate is also described as a diurnal climate. Outside the tropics, however, the term seasonal climate is used. In addition, differences in day-length in the course of the year are small. On the equator the sun rises at 6 o'clock in the morning at all times of the year and sets at the same time in the evening. As one goes further north or south of this line the days become only slightly shorter, so that in Jamaica, in the second half of December, sunrise and sunset are only half an hour later or earlier respectively.

The closer one gets to the Tropics of Cancer and Capricorn, the more noticeable are the dry periods. Precipitation may reach its lowest level on the borders of the area, in the region of the trade-winds, as is clearly shown by the desert-belts of Africa. Large parts of the tropical region are therefore conspicuous on account of a more or less long, rain-free period. Zones of vegetation vary accordingly, as may be seen from the tables on pp. 10–11. Lastly, the duration of the particular wet or dry season is of crucial importance in the cultivation of economic and ornamental plants.

Briefly it may be said that the environmental factors mentioned above led to an abundance of life which in its diversity was not surpassed in any area of similar size in the world. In the tropics the opportunity for adaptation in plants and animals is correspondingly great. This is reflected in the enormous number of species. Add to the high temperatures abundant rainfall, distributed equally throughout the year, and you have the best possible conditions for development as is shown most impressively by the tropical rain-forest. Growth and decay succeed each other so rapidly that few nutrients remain in the soil. Agricultural use of land is only possible on a limited scale.

Interference in the delicately balanced ecology of the tropics can have catastrophic effects, as we see again and again by the increase in famine conditions in the developing countries.

## Influence of Mankind on Tropical Vegetation

The influence of mankind on the environment goes back to the early history of human development. The first human interference in the land came when

| *Vegetation Zone* | *Regions* | *Climate* | *Plant Communities and Economic Plants* |
|---|---|---|---|
| Tropical Rain-forest | *America:* C. America, Caribbean Islands, Amazon region, W. coast of Colombia, Orinoco region, S.E. coast of Brazil. *Africa:* Congo region, W. coast of Ghana, Senegal, E. coast of Madagascar. *S.E. Asia:* W. coast of India, Ceylon, parts of Burma and Thailand, Malaysia, Indonesia. *Australia, Oceania:* N.E. coast of Australia, New Guinea, all tropical Pacific Islands. | 0–max. 1500m, humid hothouse climate, no seasonal changes, average annual temperatures between 24° and 30°C. Abundant rainfall spread almost equally over the whole year. Annual rainfall amounts more than 1500mm rising to 10,000mm. Relative humidity higher than 80%. No directional winds apart from monsoon winds. Formation of tropical cyclones in border regions. | Predominately trees forming dense forest, sometimes 100 different species per hectare. The tree canopy can reach heights of more than 50m. Lack of light often causes sparse development of undergrowth. The numerous lianes and epiphytic species are characteristic. Nowadays often supplanted by secondary forest and grassland. *Economic Plants:* teak, mahogany, jacaranda, cocoa, banana, rice, manioc, sugar-cane, coconut palms, oil palms. |
| Deciduous Forest and Savannah | *America:* Pacific coast of Mexico, W. coast of C. America, Caribbean coast of S. America, Llanos regions, E. and C. Brazil. *Africa:* Sahel zone, E. and S.W. Africa, W. coast of Madagascar. *S.E. Asia:* central parts of India, Burma and Thailand. | 0–about 1000m. Seasonal alternation of rainy and dry periods. Only slight variation in temperature between 18°C in coolest month and 28°C in warmest. Often considerable daily fluctuation in temperature. Annual rainfall amounts varying between 200 and 1500mm. Strong trade-winds in coastal regions, replaced in S.E. Asia by monsoon winds. | Monsoon forest, deciduous during the dry season (acacias, mimosas, *Bombax* species, baobabs) changing to sparse tree and shrub areas where it meets water-courses and savannah regions. Thorn-bush savannah in regions with more than 4 months continuous drought, in America replaced by tree-cacti and species of *Agave*. *Economic Plants:* cotton, sisal, maize, sugar-cane, manioc, citrus fruits. |

| *Vegetation Zone* | *Regions* | *Climate* | *Plant Communities and Economic Plants* |
|---|---|---|---|
| | *Australia, Oceania:* central part of New Guinea, N. and E. coast of Australia. | | |
| Semi-desert and Desert | *America:* W. and C. Mexico, W. coast of S. America, E. part of Brazil. *Africa:* Sahara, Sudan, S.W. Africa. *S.E. Asia:* central parts of India, Arabian Peninsula. *Australia, Oceania:* central parts of Australia. | 0–about 1000m. Dry climate of the trade-wind region. Beginning of seasonal variation. Average monthly temperature varying between 15° and 30°C. Rainy period often less than 1 month in the year. Annual rainfall amounts between 0 and 200mm. Very low humidity (except Peru). Considerable daily fluctuation in temperature through radiation. | Hardly any areas covered with vegetation. Scattered growth of thorn-bushes, cacti and other succulents. Cultivation of economic plants only possible with artificial irrigation. *Economic Plants:* date palms, cotton, sisal. |
| Mountain region | *America:* central parts of C. America, Andes region. *Africa:* Ethiopia, E. Africa, Mt Cameroun. *S.E. Asia:* N. India, Burma, Thailand, Borneo, Sumatra, Java. *Australia, Oceania:* N. Guinea. | 1000–1500m. Temperature reduced by altitude. Montane zone between 1000 and 2500m with average annual temperature 15°–22°C and annual rainfall more than 2000m. High mountain zone beginning at 2500m. | Montane forest as rich in species and dense as the rain-forest, strongly developed undergrowth. Epiphytic growth more frequent. Tree-ferns characteristic of this zone, in America also Bromelias and Tillandsias. Large areas infertile, due to erosion after clearance by burning. *Economic Plants:* coffee, tea, eucalyptus, maize, potatoes. |

nomadic tribes of gatherers and hunters became farmers. Then as now, the predominantly nomadic agriculture practised by some races in the primeval forest had no great influence on the environment, provided that only small groups of people were involved.

Further development in society and the state resulted in agriculture on a large scale which required a considerable amount of land. This was obtained, for the most part, by uprooting the primeval forest trees. The first catastrophic effects on the environment were already visible in pre-Christian times in the Near East and North Africa. At that time Carthage was the granary of the Roman Empire. As deforestation continued both water supplies and circulation of nutrients were disturbed. Long dry spells, broken by short periods of heavy rain, during which the nutrients were swept away by erosion, changed the once fertile regions of cultivation into the deserts of today. Uncontrolled stock-breeding, especially of goats, led to similar results by the destruction of the undergrowth and young trees. The infertile karst and maquis landscape in the mediterranean region is largely attributable to this.

Agricultural needs and subsequent effects on the environment are the same today. The continually rising population in tropical regions and modern technological opportunities for the acquisition of new agricultural areas accelerate this process. What was impossible a few decades ago – the wholesale devastation of tropical rain-forests to provide land for agriculture – is now a stark reality.

If the deforestation of the Amazon region continues at the same rate, by the turn of the century there will be nothing left of the rain-forest which previously covered an area of 8 million sq. km.

We now know that, next to the tropical coral-reef, the equatorial rain-forest is the most complex environment in the world. It is also the largest oxygen producer, whose soil, because of rapid metabolism, is characterised by an extreme dearth of nutrients. This is the reason for the lack of success in the large-scale settlement projects along the Trans-Amazon Highway in Brazil.

Recent effects of large-scale deforestation have been the creation of famine areas in the Sahel zone in Africa and in north-east Brazil. There, the predominantly dry zones with secondary forest consisting of thorn-bushes and succulent plants show the final result of human influence on tropical vegetation.

## Natural and Present Distribution of Tropical Economic and Ornamental Plants

It is significant that many economic and ornamental plants are cultivated nowadays in all tropical regions of the world. With human help they have spread from their original home throughout the whole of the tropics. In some cases their native region is known exactly, but in others their origin cannot be established. For example, we know that the rubber-producing species of the genus *Hevea* occur wild in a large area of the Amazon region. South America is also the home of the pineapple. Six different wild species have a distribution area stretching from Venezuela in the north to S. Brazil and E. Argentina in the south.

The origin of many other tropical economic plants can no longer be ascertained. The reason is that many of them have already been in cultivation for thousands of years. As new forms are continually being produced the original species have gradually died out, which is the case with our corn crops.

This is not true of all tropical economic plants as is shown by the case of the coconut palm. On the one hand there is reason to believe its home is in the region of the South Sea Islands or the Indomalayan area. On the other, there is evidence that S. America is its country of origin, for closely related species can be found there growing wild. The problem is difficult to solve because the plant produces buoyant fruits which can float great distances with the help of ocean currents. It has been proved that they can travel up to 4500km and still remain viable.

The distribution of tropical ornamental plants is even more difficult to determine since they outnumber the economic plants. Besides this, the limits of the natural distribution of decorative plants are vague. For example, within their native area, many species of orchids and bromeliads have been

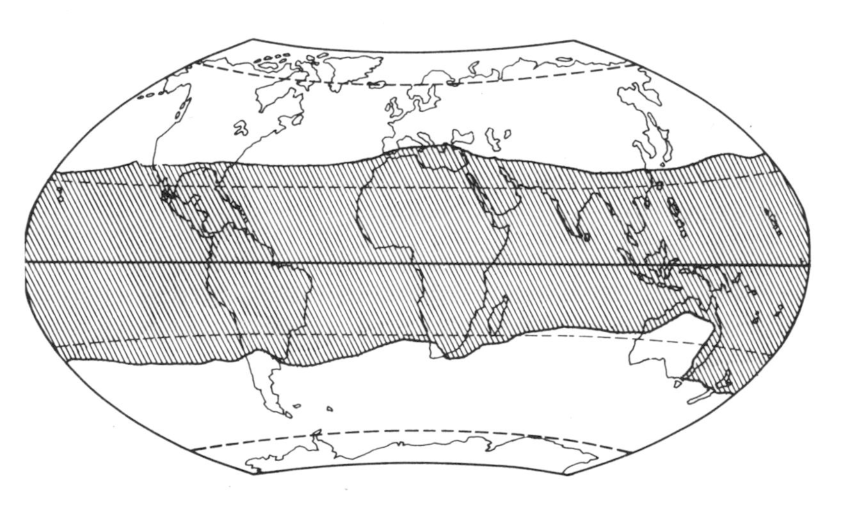

World distribution of palms (Palmae)

planted in the neighbourhood of human settlements. From there many of these very beautiful plants have been taken to all parts of the tropics. This is also true of the Bougainvillea (*Bougainvillea spectabilis*) which comes from the Amazon region. The tropical Tulip Tree (*Spathodea campanulata*), however, which has bright red, cup-shaped flowers, has its home in West Africa. The Flamboyant (*Delonix regia*) is known to originate in Madagascar, and the native area of the Royal Palm is on the islands of the West Indies. In general it may be said that the original distribution of tropical ornamental plants is known more precisely than that of many of the economic plants.

Palms form a separate group within the tropical vegetation and include many decorative species as well as some very important economic plants. They constitute a family of about 3500 species. Some of them are associated with tropical coastal plains, while others such as *Ceroxylon alpinum* are found on mountains at heights of 3000m or more. If one looks at the general distribution of the family (map, p. 13) it is clear that some species are found both north and south of the tropical region. For example the Dwarf Fan Palm, *Chamaerops humilis*, is native in S.E. Spain and S. Italy. In Japan and China the distribution of the Chusan Palm, *Trachycarpus fortunei*, extends even further north, while the Nikau Palm, *Rhopalostylis sapida*, reaches 37° south in New Zealand. Apart from these exceptions palms are a tropical group of plants with their principal centres in the Amazon region and the Indomalayan area. Their whole magical beauty is inextricably connected with the tropics.

## Domestication and Cultivation of Economic Plants

In contrast to the natural evolutionary development of plants and animals, the raising of economic plants is dependent upon man's own design and selection. Particular plants are chosen from among the species available in the wild and then these are grown and crossed with each other. Consequently domestication means that man carefully selects a group of wild plants and animals for his domestic needs, and continues the development of the selected forms by hybridisation, further selection, and in recent times even by deliberately interfering with their heredity.

The history of domestication and cultivation can be divided into five periods. From the earliest human times until the latter part of the Old Stone Age, 40,000–15,000 BC, *Homo sapiens* was a gatherer and a hunter. During this period people spent most of the day wandering about in order to find the plants and animals they needed. In the course of these wanderings they tested many of the plants they collected for their usefulness. Probably many a death was caused in this way, but also many medicinal and economic plants were discovered.

Next came the period of the harvesting peoples. Suitable wild plants were collected in large quantities and stored. Even today, in the Chaco region of S.

America for example, the pods of certain species of trees (*Prosopis alba* and *Zizyphus mistol*) which contain starch and sugar, are harvested in large quantities and put into storage, while the Indians on the upper Amazon and Orinoco still live on the starchy fruits of the palm *Bactris gasipes* which they store away for future use.

The period of gathering and storing forms the transition to the New Stone Age, when humans began to cultivate particular economic plants. They had learned how to select the seeds from certain plants and how to sow them. By finding out the right time for sowing they could give them favourable conditions for germination. They weeded the plants they were growing, and prevented them from being eaten by animals. This third period is connected with the transition from a wandering to a settled way of life, and probably occurred in the coastal regions where streams and rivers flow into the sea.

This, then, was the situation at the beginning of the fourth period, when true domestication began. Now, not only were the best seeds and hybrids chosen, but the plants obtained in this way became more and more dependent on human intervention, often to such an extent that they could no longer exist without his (or more usually her) caring and protective hand and could not compete under natural conditions. In the domestication of tropical plants it was not just a matter of how productive they were, sometimes strange forms or plants with beautiful brightly coloured seeds were chosen. It is worth noting that, in changing from the storage of plants to their domestication, similar methods independent of each other were adopted in different parts of the world.

The latest period of development, the age of planned plant culture, only began about the middle of the last century. What had previously only been experiments in this field became reality with the discovery of the principles of fertilisation in plants and the realisation of the laws of heredity. Today this is being carried further by experiments into gene mutation and the application of tissue culture.

It is interesting that the first centres of domestication and cultivation were in tropical Asia and America. Early centres of cultivation of tropical economic plants may be found by the Indus and in the Mekong delta as well as in the area of advanced Indian culture in S. America. There is evidence that a very early period of agriculture occurred in Siam. Seeds of peas and beans were found there which date back to about 9000 BC. The beginnings of the cultivation of economic plants in S.E. Asia can be traced back to 11,000–9,000 BC.

It is of great significance for the future nourishment of the world that all the potentially useful plants in Europe have been examined in detail. Without doubt alternative sources, which have not yet been exhausted, remain in the tropics. The recognition of these facts, and the realisation that the population of the world is increasing by some 75 million each year, has caused a new branch of science to appear – ethnobotany. By using the experience of native peoples, previously unknown kinds of plants could be made available to us. Time is running out, for with increasing civilisation the number of aborigines is dwindling fast and with them knowledge of the plants known only to them.

There are still many more species of palms that might be useful for providing oil. There may also be a number of medicinal plants, known only to native peoples, that could be of use to us. As such species form a larger part of the environment than highly developed economic plants, their cultivation could be the best way to protect the environment.

## Tropical Economic Plants in Mythology and Art

In the tropics, concern for daily food and the continual problems of cultivating important economic plants stimulated the human imagination. Mythological and religious ideas found their expression in painting and sculpture. In general, it is difficult to draw a line between mythology and religion, as is seen by the mixture of Christian and pagan ideas in the religious processions of Latin America or the ceremonies held by the Indians in front of the church at Chichicastenango in the Guatemala highlands. Accordingly there are numerous representations of tropical economic plants. They range from primitive drawings and ornaments to fine works of art, and often give a clue to the appearance and habit of the kinds of plants cultivated in those days.

Amongst the Indians of Latin America the tradition of maize culture has for thousands of years been expressed in their pottery, which reaches a high artistic level. We are familiar with clay vessels from Central America which are ornamented with maize cobs. From the same region we have the sculpture of a mythological maize-god whose head is similarly decorated. From these artefacts and also from nomenclatural evidence it can be deduced that the development of maize as a cultivated plant must have taken place all of a sudden. It appeared like a gift from the gods, for the chance crossing of the original maize with a closely related form resulted in a rapid and extraordinary increase in yield. The first botanical illustration of a maize plant is in a herbal by Hieronymus Bock dated 1539. There is also a very good picture of the plant in the herbal by Leonhard Fuchs published in 1542, a work famous for the special quality of its illustrations.

Not only mythological and religious stimulation, but also the sheer pleasure derived from artistic activity, can lead to the creation of works of art. For example, the round fruits of the Bottle Gourd, *Lagenaria siceraria*, from Peru, have been ornamented in an artistic way with poker-work designs. These are found not only in Peru and Ecuador but also in Mali in Africa and we must therefore assume that this form of decoration has arisen independently in the two areas. Even the dry hollow fruits of the Calabash Tree, *Crescentia alata*, can be carved and painted so tastefully that they represent real works of art. It is understandable that palms were the subject of much artistic work. Of these, the main economic and ornamental species is the Coconut Palm, *Cocos nucifera*. It is surprising that the first known artistic representation of the Coconut Palm was on a Phoenician silver bowl made 2700 years ago, for this species of palm flourishes only in the humid, tropical regions of the world. This discovery proves that the Phoenicians of the mediterranean region were already familiar with the Coconut Palm.

In economic importance, the Coconut Palm is closely followed by the Date Palm, whose most southerly point of cultivation stretches into the tropical zone. We know that this extremely important palm, which is used for a wide variety of purposes, produces both male and female flowers. It is amazing that the Assyrians were already aware of this fact nearly 3000 years ago. A highly artistic relief dating from this period shows priests in bird masks pollinating female Date Palms. It also proves that the Assyrians were already familiar with the practice of increasing the yield of their date plantations by artificial means. Nowadays it is customary for only a few male Date Palms to be cultivated in oases as sources of pollen, so that the majority of plants can be used to produce the maximum quantity of fruit. The representation of priests with birds' heads shows to what extent mythological and religious ideas can become mixed, and how they find expression in forms of art.

## Economic and Ornamental Plants on Stamps

Tropical economic and ornamental plants are often illustrated on postage stamps. As a rule the plants depicted are found wild or are cultivated in the countries on whose stamps they appear, but there are numerous exceptions. In the old days tropical plants were only found on the stamps of certain countries, but since the Second World War, and particularly in more recent times, this has all changed. A well-known example is the set showing tropical orchids which was issued in 1968 by East Germany.

Many countries often have stamps showing economic plants which are characteristic of the country concerned. In this context one may mention the set with coffee bushes issued in 1940 by the Central American republic of El Salvador, whose economic existence depends to a large extent on the cultivation and export of coffee. Other tropical countries have illustrations on their stamps of tropical fruits which play an important part in feeding the native population. For example, a set of stamps was issued in 1968 by Nicaragua which include the mango, pineapple, orange, pawpaw, banana, avocado pear, water-melon, cashew nut, sapote and cocoa. A very beautiful stamp from the Philippines shows the fruit of the Carambola (*Averrhoa carambola*), Sugar Apple (*Annona squamosa*), Mangosteen (*Garcinia mangostana*) and Paprika (*Capsicum annuum*) with a banana as the background. These are just two examples from the New and the Old World.

Although palms occur less frequently in the design of stamps, there are some interesting examples. During the period 1891–1915 the republic of Haiti issued both sets and individual stamps bearing more or less stylised representations of the Royal Palm, *Roystonea regia*, whose natural area of distribution covers the West Indian islands. In 1902 the neighbouring island of Cuba showed, on one of the first stamps to be issued after gaining its independence, the same palm which has great significance for the island as a decorative and economic plant. There it is rigorously protected, particularly in W. Cuba where it is the only kind of tree to be planted after the forest has been cleared to provide more land for cultivation. Other stamps showing palms have been issued by Brazil, Colombia, Honduras, El Salvador, Malawi, Niger, North Borneo, Indonesia and Oceania. There are also interesting stamps from French colonial times which illustrate the uses of tropical palms. For example, stamps from Dahomey in West Africa, which were issued during the period 1913–17, show natives harvesting the fruits of the Oil Palm, *Elaeis guineensis*. In the same period there were stamps from Cameroun showing how rubber is obtained.

So many tropical ornamental plants have appeared in stamp design that it is almost impossible to summarise them. Orchids, which include some of the most decorative tropical plants, have been featured particularly frequently. One example from the genus *Cattleya* may be mentioned, *C. trianae*, which is depicted on Colombian stamps. Even in temperate latitudes the large-flowered *Cattleya* species are favourite ornamental plants, as is shown by the set from East Germany mentioned above. Stamps with tropical orchids have also been issued by Venezuela, British Honduras, Cuba, Jamaica, Barbados, Burundi, as well as Papua-New Guinea and many other countries. Finally, there is a stamp from Panama showing the national flower of the country, a kind of orchid known as the Flower of the Holy Ghost.

## Growth Forms

The terms tree, shrub and herbaceous perennial are used to describe the growth form of a particular plant. These categories are determined by the position of the organs of perennation, that is the buds, during the unfavourable season of the year. In herbaceous perennials these are situated directly above the surface of the ground and are generally protected by the remains of the leaves. In shrubs and trees on the other hand the buds occur higher above the ground, and in tropical plants the protective scales are often absent. Like trees and shrubs, herbaceous perennials can live for many years.

In contrast to herbaceous plants, the main and side shoots of trees and shrubs become woody. While trees can reach heights of 100m and more, shrubs remain much lower. The two forms can often be distinguished purely on height, for shrubs do not usually grow more than 5m high. But in the

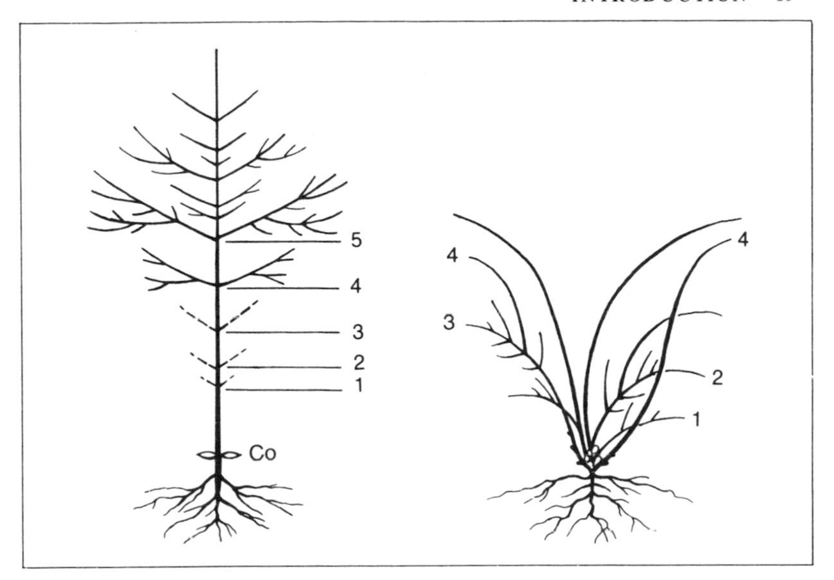


**Branching of trees and shrubs**
In trees a main axis is always present, but in shrubs the original main axis is outstripped by the lateral shoots and pushed to one side. 1 = main shoot; 2, 3, 4 etc. = lateral shoots; Co° = cotyledons.

tropics a wide range of intermediate forms is found, and it is often difficult to tell if a woody plant is a tree or a shrub.

However, the two groups can be clearly separated by their morphological structure. In trees there is always a main axis or trunk from which the lateral axes or branches arise. But in shrubs, even in the early stages of development, there is obvious growth at the base of the lateral axes. They rapidly outstrip the main axis which either remains at the centre or is pushed to one side (see diagram). The equal development of numerous side shoots at the base leads to typical shrub-like growth. In many tropical shrubs the shoot contains a distinct, non-woody pith. In a narrower sense we can recognise dwarf shrubs. They also have woody shoots, but are distinguished by their lower growth. In the tropics they are less important.

# Ornamental Plants

## Tropical Ornamental Trees

### Flamboyant, Flame Tree, Peacock Flower, Royal Poinciana 1
*Delonix regia* (Boj. ex Hook.) Raf.

**Senna family** Caesalpiniaceae

The Flamboyant is one of the most magnificent ornamental trees of the tropics and is native in Madagascar. It reaches 10–15m in height and at maturity is particularly noticeable for its broad, umbrella-shaped crown. Its doubly pinnate leaves, composed of numerous tiny leaflets, reach a length of 30–50cm and are shed during the dry season. The scarlet flowers appear in spring, before the leaves develop, and have orange-striped petals which narrow considerably towards the base. They are the national flower of Puerto Rico, and are pollinated by birds. The brown pods, which may be flat or wrinkled when ripe, grow up to 60cm long and are especially striking. The tree is often planted to provide shade in parks, gardens and avenues. Its fine pinnate leaves form a wonderfully delicate mosaic against the blue tropical sky. The wood is soft, and the trunk exudes resin. The numerous leaflets fold up quickly on the approach of dusk. This phenomenon can be observed in many trees with papilionaceous flowers.

### African Tulip Tree 2, 3
*Spathodea campanulata* Beauv.

**Trumpet-flower family** Bignoniaceae

This tree was first described in 1787 and is one of the most striking of tropical ornamental trees. It is planted in the country as well as in built-up areas, and its large inflorescences are visible from a distance. The tree reaches a height of 10–15m and is usually of erect growth. It is native in tropical West Africa. It has pinnate leaves, shiny above, and 30–40cm long. The large cup-shaped flowers are in dense clusters and the outer ones open before the inner. They are bright red in colour with a decorative gold margin. Water collects within the unopened flower-buds, and consequently the inflorescences used to be described as 'water-cups'. The flowers are followed by erect brown capsules, some 20cm in length, which split along the side to release numerous seeds with a transparent winged edge. This genus contains only two species, both from tropical Africa.

### Jacaranda 4, 5
*Jacaranda mimosifolia* D. Don

**Trumpet-flower family** Bignoniaceae

The Jacaranda is also known as the Fern Tree, a name which comes from its Latin synonym *Jacaranda filicifolia* (fern-leaved). Botanically, however, this name is not very appropriate. It is a valuable ornamental and economic tree which reaches a height of up to 10m. Its finely divided, doubly pinnate leaves, which are shed during the dry season, contrast with the smooth, grey bark and are as delicately ornamental as those of the Flamboyant. The tree is native in Brazil, where it grows in savannah-like plant communities.

The delicate, bell-shaped, blue-violet flowers open shortly before the leaves, and are arranged in paniculate inflorescences. They form a colour combination of rare aesthetic charm against the azure blue of the tropical sky. After the two-celled ovaries have been fertilised, round woody capsules develop which are about 8cm in diameter and may be slightly wavy-edged. The genus contains 40 species from tropical America and the Antilles. Several species are sources of a soft wood which is used for making furniture and for carving.

## Golden Trumpet-tree 6

*Tabebuia chrysantha* (Jacq.) Urban

**Trumpet-flower family** Bignoniaceae

The genus *Tabebuia* contains about 100 species which are distributed from Mexico and the West Indies to N. Argentina, and the Golden Trumpet-tree from Central America is one of the most spectacular. At the beginning of the dry season it sheds its leaves and is then covered all over with golden yellow, bell-shaped flowers. It grows to about 15m in height and has undivided, leathery leaves arranged in pairs.

The tree produces hard wood of good quality that is used for making railway sleepers and cart wheels and in turnery. Because of its slightly yellowish colour it is also called Yellow-wood (cortez amarillo). It contains the dye lapachol and is highly resistant to fungal attacks. The wood of other species of *Tabebuia* is known as ipé-wood. *T. flavescens* is the source of green ebony.

## Pink Trumpet-tree 7

*Tabebuia rosea* (Bertol.) DC.

**Trumpet-flower family** Bignoniaceae

The Pink Trumpet-tree, like the Golden Trumpet-tree, is native in Central America and is also planted in parks, gardens and avenues. It may reach 15m in height but is usually lower. This tree has been chosen as the national plant of El Salvador where it occurs in seasonal deciduous forest. Its clusters of pale to deep pink flowers are very decorative and visible from afar. It too flowers in a leafless state at the beginning of the dry period from October to December. Like the Golden Trumpet-tree it produces valuable timber, pale grey in colour with purplish veins, which is used in the manufacture of furniture. The freshly cut wood becomes darker as it dries, without losing its highly ornamental grain. The tree is also used for decorative purposes as a shade-tree in coffee and cocoa plantations.

## Tropical Fig Trees 10, 11

*Ficus* spp.

**Mulberry family** *Moraceae*

Many of the 100 or so species of the genus *Ficus* are grown as ornamental trees in the tropics, particularly as shade trees, for they often have a broad flat crown or form an arch over streets and avenues. The blade of the evergreen leaves has a rounded base and is at times drawn out into a distinct point which leads water off the leaf when it rains. For this reason it is termed a 'drip-tip'.

The fruits as a rule resemble those of our edible fig, and develop from the urn-shaped inflorescence formed from the axis of the shoot. Inside, numerous male and female flowers are packed tightly together. As it ripens the fruit-wall usually becomes fleshy, and is often eaten by animals, e.g. bats. The colour of the fruit can be green, greenish yellow, red, purple or dark violet, and its size varies from a few millimetres to several centimetres. The actual fruits which are found inside the fleshy inflorescence when it is ripe are like tiny nuts.

Many *Ficus* species are known to strangle trees. Seeds are brought, usually by birds, to a fork in the branches or the niche where an epiphyte is growing, and there they germinate. The young plant forms numerous roots which try to reach the ground. Consequently, the original host-plant can be entirely covered by the net-like lattice formed by the roots of the *Ficus* plant, until it finally dies and disintegrates, leaving behind the bare network of adventitious roots. This kind of root-formation, which is characteristic of the Rubber Plant, is also typical of *F. dealbata*. Lateral roots connect with the vertical ones, and tufts of thin hairs arise which take up water and nutrients.

In the case of a number of *Ficus* species, the formation of adventitious roots means that the trees stand on hundreds of thick columns, and a single tree can extend over a large area. The Banyan, *F. bengalensis*, from India is a well-known example of this. The species *F. aoa*, which is found on the islands of Samoa, produces some of the mightiest trees, and the leaves of *F. religiosa* have particularly noticeable drip-tips. According to legend Buddha received his enlightenment under a fig-tree in 500 BC. Since then, Buddhists have regarded it as sacred. A descendent of this tree at Anuradhapura in Ceylon is reputed to be over 2000 years old.

A favourite ornamental plant is *F. lyrata*, a small tree native in tropical W. Africa, which has thick, leathery, dark green leaves 50–60cm in length. They are broadly obovate in shape, slightly rounded at the tip and remarkable for their lyre-shaped base. The large, raised leaf-veins are especially conspicuous.

## Rubber Plant

**8, 9**

*Ficus elastica* Roxb.

**Mulberry family** Moraceae

The Rubber Plant is one of the most familiar of tropical house-plants, but it should not be confused with the Rubber Tree which belongs to a quite different family. Its importance as a source of rubber has declined considerably. It is native in S.E. Asia, where it occurs as a large tree in the virgin forests of Assam and Burma to Sumatra and Java. The Rubber Plant reaches a height of more than 30m and has alternate, broadly lanceolate leaves which are widened at the base and drawn out to a point at the apex. In bud they are enclosed in a membranous, bright red stipule which falls away as the true leaves unfold.

Like other *Ficus* species, the Rubber Plant bears urn-shaped inflorescences. Its fruits are relatively small and narrowly elliptic in cross-section. It too forms adventitious roots on the lower branches which grow down to the ground. When they have done so, they quickly thicken and may become branched.

In earlier times, the Rubber Plant was cultivated as a source of rubber, especially in S.E. Asia and W. Africa. The whole plant contains latex canals, which exude a white fluid when cuts are made in the trunk. But today its significance as a source of rubber has declined in favour of the Brazilian species of *Hevea*.

## Yellow Trumpet-tree, Yellowbells, Yellow Elder, Yellow Bignonia 12

*Tecoma stans* (L.) H.B.K.

**Trumpet-flower family** Bignoniaceae

There are 16 species in the genus *Tecoma*, distributed from Mexico and tropical America to the Antilles. Among these is the Trumpet Vine, *T. radicans*, a bright orange-flowered climber which may be grown outside in the milder regions of Europe. The Yellow Trumpet-tree on the other hand forms a small tree or shrub up to 9m in height. It is very popular as an ornamental plant and widely cultivated. Its pinnate leaves are up to 30cm long, and are composed of 5–13 long-pointed leaflets 4–10cm in length. After the bright yellow petals have fallen, two-celled capsules develop, similar to those of the *Tabebuia*. The wood of this little tree is used for building purposes, and also as firewood. A closely related species, *T. leucoxylon*, provides the green, brown or yellow ebony of the Antilles, which because of its variable colouring is also called bastard guajak. It is used for fine joinery and turnery.

## Frangipani, Nosegay 13

*Plumeria alba* Griseb. non L.

**Dogbane family** Apocynaceae

The Frangipani has numerous English and Spanish names, which shows that this striking and familiar plant is widespread in the tropics. Its original home is the West Indies. It is a small tree or shrub which is conspicuous on account of the regular manner in which the branches fork. It has strikingly thick, smooth, green twigs which ooze large quantities of latex if they are damaged. The linear-lanceolate leaves are glossy on the upper surface and pointed, and reach a length of 20–50cm. They fall off at the beginning of the dry season leaving behind large scars on the branches. The flowers are strongly scented and the arrangement of the petals resembles the blades of a fan. The flowers are white with a yellow mark at the base. The twisted appearance of the flower-buds is typical of the Apocynaceae. The fertilised flower develops into two conspicuous, narrow pods, up to 25cm long, which are joined together at the base. The tree is named after the French botanist Charles Plumier, who travelled the Caribbean area in the seventeenth century. It is often planted near temples and in cemeteries in Ceylon, India and S.E. Asia. The pink-flowered Frangipani from Central America, *P. rubra*, is closely related, but is distinguished by its obovate leaves which are up to 45cm long, and less hairy on the underside.

## Mahoe, Cuban Bast 14

*Hibiscus elatus* Sw.

**Mallow family** Malvaceae

The Mahoe is best distinguished from the common and widespread Sea Hibiscus by its growth and greater height. The Sea Hibiscus bears at the ends of the shoots clusters of two or three flowers which are lemon yellow in the morning and change to orange-red and then dark red in the course of the day. The Mahoe is native in the deciduous forests of Cuba and Jamaica, and the relatively narrow petals do not overlap each other to the same extent as those of the Sea Hibiscus. The flower is orange when it opens in the morning but becomes a fiery red as the day progresses. The tree flowers throughout the whole year, especially during the winter months. Like the Sea Hibiscus its fruits are in the form of globular capsules, but its wood is somewhat darker and harder, and is used principally in the manufacture of furniture for building purposes. The fibrous bark is used for cordage (rope for rigging) like that of the Sea Hibiscus. A tea to alleviate dysentery is prepared from the shoots.

## Indian Almond

*Terminalia catappa* L.

15, 16

**Combretum family** Combretaceae

The Indian Almond is found in almost all parts of the tropics. Because of its tolerance of salt, it is mainly cultivated as a shade-tree in coastal regions. Its natural distribution stretches from Madagascar, through India, to the Fiji, Ryukyu and Bonin Islands. In its young state, the tree may be recognised by its horizontal branches which grow in circles at different levels on the trunk. These bear clusters of very large, elliptic leaves which are pointed on young trees. The characteristic arrangement of the branches disappears with age, and the mature tree, which grows up to 10m high, then has a broad, spreading crown.

The slightly flattened, narrowly two-winged fruits are 5–6cm long and possess a layer of fibrous tissue between the outer, green, fleshy shell and the hard, woody nut. These fruits, which will float in water, enable the Indian Almond to spread far and wide, like the Coconut. It is often found on sandy shores where it germinates and sends down long roots. The discarded fruits are also frequently found under the trees in streets and parks, and are highly prized for their seeds which have an almond-like taste.

Before the leathery leaves fall, they turn a bright red. Both the shell of the fruit, which is rich in tannin, and the bark of the trunk are used in tanning, and in India a black dye is also made from them. The genus comprises roughly 200 species distributed from South and East Africa, through Indonesia to N. Australia and Polynesia.

## Horse-tail Casuarina, Australian Beefwood, Australian Pine, South Sea Ironwood

*Casuarina equisetifolia* L.

17, 18

**Casuarina family** Casuarinaceae

The Horse-tail Casuarina can be found nowadays in almost all coastal regions in the tropics, though its original distribution extended from N. Australia and Indonesia to Africa. The tree has an odd appearance because of the numerous green, finely divided branches, which form a delicate filigree-work. On these finely grooved branches appear whorls of leaves, reduced to scale-like teeth. The individual slender shoots resemble the plant known as the Horse-tail, which accounts for the Latin name of the species (*Equisetum*=Horse-tail). Both the female flowers, which have red stigmas, and the male flowers are inconspicuous and are crowded together at the ends of the branches. The female catkins develop into clusters of fruit as large as a hazel-nut, composed of numerous, small, woody fruits arranged in whorls. They split open when ripe, releasing a winged seed.

The tree reaches over 20m in height and is used for afforestation in the tropics. It is salt-tolerant, quick-growing and undemanding. Its wood is hard, heavy, and difficult to cut, but is good for burning. Because of its hardness it is classed as an ironwood.

The genus *Casuarina*, which contains about 50 species distributed mainly in Australia and New Caledonia, occupies an isolated position in the plant kingdom.

## Screw Pine, Screw Palm — 19, 20

*Pandanus* spp.

**Screw Pine family** Pandanaceae

The Screw Pine owes its name to the spiral arrangement of its leaves. Approximately 630 species are known, most of them occurring in forests. A few are coastal plants, tolerant of salt, and these are often found as ornamental plants near beaches. The natural distribution of the genus extends from Africa and Madagascar, through S.E. Asia and Australia to the Pacific Islands. Apart from the screw-like arrangement of the sharply toothed, linear, long-pointed leaves, Screw Pines are usually recognisable by the adventitious roots at the base of the trunk. These may arise several metres above the ground and often hang down stiffly like aerial roots. At an advanced age the trees bear globular inflorescences as big as one's head, composed of numerous individual fruits arranged in the form of a polyhedron.

Quite a number of species are used in various ways. The tough leaves are used for making sleeping-mats, baskets, hats, screens, and for covering huts. The leaf-fibres are employed in all kinds of plait-work, including the manufacture of ropes, nets and belts. The most important fibre-plant is *P. utilis* from Madagascar (Photo 20). It is cultivated mainly in the West Indies, is distinctly screw-like in form, and has dark green leaves, up to 1.5m long, with firm spines. The male inflorescences of the Hala Screw Pine, *P. odoratissimus*, are highly regarded because of their pleasant scent and are used by Polynesian women to adorn their hair. That is why this species is often cultivated there. In Madagascar the fleshy part of the fruit of some species, e.g. *P. edulis*, is eaten, and on the Marshall Islands a variety tasting like apples, *P. odoratissimus* var. *laevis* has long been cultivated as a fruit tree. The wood is used in building houses, especially on the coral islands of the Pacific, where timber is scarce.

## Silk Oak, Silver Oak — 21, 22

*Grevillea robusta* A. Cunn. ex R. Br.

**Protea family** Proteaceae

The Silk Oak is one of the most popular of tropical ornamental trees, and is also grown as a shade tree in those coffee and tea plantations situated at higher altitudes. It has spread from its home in E. Australia over the whole of the tropical zone. It reaches a height of 25–50m and its leaves are divided almost to the midrib into narrow lobes with silvery hairs forming a pelt on the underside. The open kind of foliage, allowing free circulation of air between the leaves, makes it suited to very sunny places.

The beautiful bright golden yellow flowers are so arranged that the inflorescence resembles a comb or a large toothbrush. Like all members of the Protea family, the individual flowers have four petals and stamens and a one-celled ovary. The petals remain joined together for a long time and the anthers are attached to them. The long style projects from the side of the flower-tube in an obvious curve. The petals separate only when old, and the extended style with its club-shaped stigma then moves into a vertical position, a procedure typical of many Proteaceae. The beautifully mottled wood of the tree is suitable for joinery. In temperate regions the Silk Oak is grown in pots as a houseplant and can easily be mistaken for a fern.

The genus *Grevillea* contains 120 species, distributed from Australia and Tasmania to New Caledonia and Melanesia. The Proteaceae are generally to be found on the southern tips of all the continents in the Southern Hemisphere. At a superficial glance they may appear to vary greatly, but the flower structure is always the same. The fruits of many members of the Proteaceae only open after fire has swept through the place where they are growing.

## Sea Grape

23

*Coccoloba uvifera* L.

**Buckwheat family** Polygonaceae

The Sea Grape forms a broad shrub or tree up to 15m in height. It originated in the coastal regions of tropical America, but because of its tolerance of salt and resistance to wind it is grown everywhere on tropical coasts. To some degree it can even stand the effects of salt-spray and foam.

The branches of this low tree bear large, very thick, round to broadly kidney-shaped leaves. In their young state they are coloured red because of the formation of anthocyanin, and they have dark-coloured dots on both sides. The inconspicuous, yellowish white flowers are arranged in hanging inflorescences 15–20cm long, and are unisexual. The female flowers are solitary, while the male are in groups of up to seven flowers. A berry-like fruit develops from the female flowers and the petals form part of its structure. It is about 1cm long, rather like a small pear in shape, and reddish purple in colour when ripe. The fruits are arranged in fairly dense clusters like bunches of grapes, hence the name of the species (*uvifera* = bearing grapes). They are edible and have a somewhat acid taste. Decoctions of roots and bark are used in cases of diarrhoea.

The genus *Coccoloba* contains about 125 species, all from the tropical and subtropical parts of the New World. Some species have large leaves, more than 40cm long and 20–30cm broad. They are often grown in glasshouses in temperate regions and make attractive ornamental plants.

## Jerusalem Thorn

24, 25

*Parkinsonia aculeata* L.

**Senna family** Caesalpiniaceae

The Jerusalem Thorn is a shrub or small tree up to 9m in height and was named after the English botanist John Parkinson. It is native in the tropics of the New World, and is cultivated in parks and gardens especially in the West Indian islands. The light and delicate appearance of the plant is due to its foliage which forms a fine tracery. The tree has long, usually pendent pinnate leaves, with a winged midrib and numerous small leaflets which are sometimes completely absent. At the base of the leaves are two stipules which have hardened into thorns (*aculeatus* = thorny). The leaflets, like those of the Flamboyant, close up as dusk approaches.

The flowers are bright sulphur yellow with orange stamens, followed later by a leathery pod, 5–8cm long, which is convex on both sides and constricted between the seeds. Decoctions of bark, leaves, flowers, and green twigs are used to reduce fever. The genus *Parkinsonia* has only two species, one of them native in Africa.

## Pink Shower 26

*Cassia grandis* L.

**Senna family** Caesalpiniaceae

The Pink Shower is one of the most striking ornamental trees of the New World tropics. It is remarkable not only for its beautiful flowers but also for its large fruits. It forms a tree up to 15m high with a broad crown, and its original distribution was from Mexico, through Central America to the northern parts of S. America and in the Greater Antilles. The tree has a smooth, grey bark, and in the rainy season bears pinnate leaves with 8–20 pairs of leaflets. During the dry season it loses its leaves and is then covered all over with pink flowers which make it visible from afar.

Like all members of the Caesalpiniaceae the flowers are papilionaceous, although the standard petal is relatively small and the upper edge of the keel petals overlaps the lower edge of the wings. The single carpel develops into an almost cylindrical, woody, dark brown pod, about 90cm long and 6–7cm in diameter. These pods can be seen from a distance as they hang down from the crown of the tree. Inside are numerous, flat, circular seeds embedded in a bitter, unpleasant smelling pulp. The tree is the source of a wood which is hard and fine-grained but not very durable.

## Golden Shower, Indian Laburnum 27

*Cassia fistula* L.

**Senna family** Caesalpiniaceae

The Golden Shower is a tree up to 9m in height with a usually broad, open crown. It is native in S. Asia, but is cultivated also in Africa and tropical America. It is a deciduous tree which, however, is never quite leafless. The change of foliage on individual branches takes place only gradually and in S.E. Asia the process lasts nine or ten months. The handsome leaves reach a length of 40–50cm and consist of ovate, pointed leaflets 7–20cm long. The splendid flower-clusters attain a length of 80cm and are composed of flowers which vary in colour from pale yellow to bright gold. The single carpel develops into a pendent, cylindrical pod up to 50cm long containing a dark brown, sweet pulp which has a laxative effect. In India the bark of the tree is used for its tannin.

The genus *Cassia* comprises approximately 500 species, some of which contain laxative substances in their leaves, others in their fruit-pulp. A few of them have leaves which provide well-known drugs. The most important species are *C. acutifolia*, Alexandrian Senna, and *C. angustifolia*, Tinnevelly Senna, which are distributed from Egypt across Arabia to India, and are sources of senna leaves.

## Amherstia 28

*Amherstia nobilis* Wall.

**Senna family** Caesalpiniaceae

The Amherstia comes from Burma and received its generic name in honour of Countess Amherst, wife of a Governor-General of India. It is one of the most beautiful of tropical trees, attaining a height of 15–18m and bearing large, pinnate, pendent leaves. The flowers have no scent but form lovely dangling clusters, narrowly pyramidal in shape. The individual flowers are about 20cm long and 12cm broad, each on a bright red stalk bearing two large bracts. The flower has four red sepals and three petals which are also bright red but tipped with yellow. The central petal is considerably enlarged. Of the ten stamens present, the five large outer ones and four of the smaller, inner ones are joined together to form a tube. The tree rarely produces seeds and requires conditions of high humidity.

## Coral Tree 29, 31

*Erythrina* spp.

**Pea family** Fabaceae

The genus *Erythrind* contains about 100 species distributed in the tropics of the Old and New World. They are trees which have a soft wood and leaves composed of three leaflets. Most of the species are deciduous in the dry season. Some of them are planted in coffee plantations as shade-trees, others are favourite ornamental plants. The most widespread species is *E. crista-galli* from Brazil which has broad, coral-red flowers (Photo 29), while *E. bogotensis* (Photo 31) has narrow flowers, directed downwards. The large, fleshy blooms are usually clustered in axillary or terminal racemes, those of some species being eaten as a vegetable in Central America. The young leaves are also used as a vegetable. The Common Coral Bean, *E. corallodendron*, produces red seeds which are made into necklaces. *E. rubrinervia* from Central America also has seeds of a fiery red colour. The seeds of some species contain substances that stupefy fish, and they are therefore used in fishing.

## Geiger Tree 30

*Cordia sebestena* L.

**Borage family** Boraginaceae

The genus *Cordia* was named in honour of the German botanist Enricius Cordus and his son Valerius Cordus, and comprises roughly 250 species of trees and shrubs in the tropics and subtropics. The Geiger Tree, which is native in the Caribbean region, is a shrub or small tree with large, stalked, broadly elliptic, hairy leaves. The brilliant orange-red flowers are clustered at the ends of the shoots and the inner ones open first. The individual flowers have a green, five-lobed calyx and a corolla with a narrow tube and spreading lobes. The two-carpelled ovary develops into a sticky, sweet fruit which can be used for treating coughs. The tree flourishes particularly well in the dry climate of Curaçao, Aruba, Barbados and Antigua. The closely related *C. myxa* also produces fruits which are used for coughs.

## Cannonball Tree 32, 33

*Couroupita guaianensis* Aubl.

**Brazil Nut family** Lecythidaceae

The Cannonball Tree is a botanical curiosity from the New World. Its natural area of distribution is in the tropical rain-forests of French Guiana and in the Amazon region westwards to Venezuela. The tree is remarkable for the frequency with which it changes its foliage, and for the position and form of its flowers and fruits. This tall tree bears alternate, elliptic, pointed leaves up to 24cm in length, which form clusters at the ends of the branches. At first the pinkish red flowers grow directly from the grey trunk, later they appear on short shoots which grow longer and become branched as the years go by (Photo 32). The flowers have six fleshy petals, and in the centre there is a helmet-shaped structure, folded inwards, formed from the numerous stamens which have joined together. At night the flowers are highly fragrant. The ovary develops into a hard, globular, reddish brown fruit at least 20cm in diameter, *i.e.* as large as a human head. Inside are numerous seeds, embedded in a cheese-like pulp with an unpleasant odour. The fruit often takes more than a year to ripen.

Changes of foliage occur several times a year in rapid bursts of growth, and on the trunks of the older, mature trees hundreds of inflorescences shoot forth, from the base to the crown, giving them the appearance of columns decorated for a festival.

## Horseradish Tree **34**

*Moringa oleifera* Lam.

**Horseradish Tree family** Moringaceae

The Horseradish Tree is native in East Africa and South Asia but nowadays it is cultivated in the tropics of the New World. It is a tree up to 10m high which resembles a Robinia but has bipinnate leaves. The paniculate inflorescences consist of yellowish white flowers with five petals. The flower develops into a bluntly triangular, brown, long and narrow, pendent capsule which reaches a length of 20–45cm. Inside the capsule are the three-winged seeds, neatly arranged in three rows. The tree gets its name from the horseradish-like taste of its roots. Its leaves and unripe fruits taste like cress, and are used as a vegetable. The oil which is obtained from the seeds is employed in the treatment of skin diseases. The juice extracted from its roots and leaves is used in poultices for inflammation and swellings of the neck.

## Sausage Tree **35, 36**

*Kigelia africana* (Lam.) Benth.

**Trumpet-flower family** Bignoniaceae

Because of its unusual fruits, the Sausage Tree has been spread by man throughout the tropics, but its home is in tropical West Africa. It is a tree up to 15m high with a broad, spreading crown and handsome pinnate leaves composed of eight to ten large, obovate leaflets. The flesh-coloured flowers, dark brownish red inside, hang from the crown of the tree in loose panicles (Photo 35). They open only for one night, giving out an unpleasant odour, and fall off in the morning. The one-celled ovary formed from two united carpels develops into a sausage-shaped fruit which has a woody stalk and a hard rind. It reaches a length of 35–60cm, a diameter of 18cm, and weighs 5–7kg (Photo 36). In its native region it is used externally for all kinds of medicinal purposes. The genus comprises ten species in tropical Africa and Madagascar.

## Floss-silk Tree **37**

*Chorisia speciosa* St. Hil.

**Cotton-tree family** Bombacaceae

The Floss-silk Tree is a quick-growing tree that reaches a height of 12–15m. Its natural distribution is from S. Brazil to Argentina. It is conspicuous because of its smooth, green, slightly barrel-shaped trunk, the surface of which is densely covered with spines. These may disappear with age. The young leaves are pale green, palmately five- to seven-lobed and stalked. After the leaves fall, in March and April, flowering time begins. The flowers are large and have five petals. These are usually yellowish white at the base, tinged purple at the apex, and often have brownish spots in the central area. There are some forms with almost entirely red flowers and others with flowers of delicate purplish pink. The tree is at its most splendid in August when the young leaves appear. It is then that the large, shining, five-valved capsules change from green to brownish green, split open and, after the separate parts of the capsule wall have fallen away, reveal ovoid masses of glossy, pure white, woolly hairs. These hang among the newly-opened leaves and resemble iridescent balloons, in which are embedded dark brown seeds, as big as peas. The genus varies widely in form and is systematically very diverse.

## Chinaberry, Bead Tree — 38

*Melia azedarach* L.

**Mahogany family** Meliaceae

The Chinaberry is one of the most widespread and most popular ornamental trees in the tropics and subtropics. It comes from the southern margin of the Himalaya, and has been planted since olden times near temples in Persia, Ceylon and Malaysia. This quick-growing tree reaches a height of 12m and regenerates rapidly by self-sown seedlings. Its doubly pinnate leaves, somewhat resembling those of an ash, form an open kind of foliage, and at flowering time the panicles of pale purple, lilac-scented flowers add to the beauty of this attractive tree. The effect is further enhanced by a dark reddish purple staminal tube at the centre of the five-petalled flowers. The three- to six-celled ovary develops into a golden yellow berry, which is why it is also known as the Bead Tree. The plant is widely appreciated as is shown by the numerous common names such as Persian Lilac, Indian Lilac, Pride of India and Paradise Tree. The Chinaberry is often planted in avenues and as a shade-tree. In El Salvador it is grown in coffee-plantations to provide shade. The wood is reddish at the centre and is used in making furniture and musical instruments. The tree contains a green dye in its leaves, and is also a source of gum-resin, oil, and various substances employed for medicinal purposes.

## Flame Tree, Flame Bottle-tree — 39

*Brachychiton acerifolium* (A. Cunn. ex Don) F. Muell.

**Cocoa family** Sterculiaceae

The Flame Tree is related to the Cocoa Tree, and is one of the eleven species comprising the genus *Brachychiton*, all of which are native in Australia. At flowering time the tree is conspicuous even at a distance for it is then that its branches are covered with small, fiery red, bell-shaped flowers. Some branches may still have retained their lovely, lobed leaves which are shaped like fans. The flowers conceal numerous individual carpels which develop into dry, unwinged fruits. These split along one side when ripe like many of the Sterculiaceae. The name *Brachychiton*, meaning 'short tunic', was given to the genus because of the shortly bell-shaped flowers. A number of species have thickened trunks, a feature of the ecologically and morphologically distinct group of bottle-trees.

## Queensland Umbrella Tree 40, 41

*Schefflera actinophylla* (Endl.) Harms

**Ivy family** Araliaceae

Because of the curious structure of its inflorescences, which are in fact compound umbels, the Queensland Umbrella Tree is also called the Octopus Tree. This decorative tree, 8–10m high, with a usually slender trunk, originates from Australia, and is easily recognisable, even in a flowerless state, by the distinctive form of its large, digitate leaves. The leaf-divisions are a glossy, dark green, linear to lanceolate in shape, and hang down from the end of a long stalk, giving rise to the name Umbrella Tree. The tree needs full sunlight to produce its flowers and it is then that the strange-looking inflorescences can be seen. The flowers are clustered on long stalks and are greenish yellow at first, but later change to pale pink and finally dark red. After the dark purple capsules have formed, the partial inflorescences fall off without disintegrating. The plant is also cultivated in glasshouses. Many species of this genus are climbers, often supporting themselves with the aid of adventitious roots. They are especially well represented in the tropics of the New World and in S.E. Asia.

## Crape Myrtle 42

*Lagerstroemia indica* L.

**Loosestrife family** Lythraceae

The Crape Myrtle is a small tree, 6–8m in height, with a smooth, grey bark, and is recognisable by its rod-like branches which are directed upwards. Its natural distribution is in E. Asia, although it is cultivated nowadays everywhere in the tropics and in the subtropics too. The leaves are relatively thick, narrowly ovate and arranged in pairs. The tree blooms in all its splendour in June, and has been called the Queen of Flowers. The lovely purplish pink flowers form dense panicles at the ends of the branches. The petals are narrowed towards the base and have a curly, fringed margin, and the stamens are numerous. There are 30 species in the genus, including the lovely *L. speciosa*, which grows to 24m high in moist riverside regions in Assam, Burma and Ceylon, and is one of the most beautiful flowering trees of S. Asia. The hard wood is highly prized, and the bark and leaves are used as a diuretic.

## Morning Glory Tree 43

*Ipomoea arborescens* (Humb. & Bonpl.) Don

**Convolvulus family** Convolvulaceae

Although it is unusual, there are shrubs and even trees in the Convolvulaceae. One of these is the Morning Glory Tree, which can reach a height of 10m and is, in fact, the largest member of this group. It is native in Guatemala, Honduras, and El Salvador, has an open crown, and is cultivated as an ornamental tree along roadsides. The flowers are 7–8cm across, snow white, with a conspicuous purplish brown throat, and appear during the dry season from November to March when the tree has shed its leaves. It thrives especially in dry places near the coast. At first the wood has a soft consistency, but it very quickly becomes hard as it dries. When struck, it gives out a metallic sound and is therefore used in making musical instruments like the marimba. There are 400 species in the genus, including *I. murucoides*, which is also tree-like but smaller then *I. arborescens*, and which originates in Mexico.

## Eucalyptus

44

*Eucalyptus* spp.

**Myrtle family** Myrtaceae

Members of the genus *Eucalyptus* are cultivated in all tropical and subtropical regions of the earth. This is especially true of *E. globulus*, a quick-growing tree with hard wood. The bark comes away in strips from the trunk, and like many other species, the evergreen leaves are pendulous and sickle-shaped. In addition to these sickle-shaped leaves, broad ovate leaves may be seen, especially on young trees. The formation of two kinds of foliage is typical of many species of *Eucalyptus* trees.

The flowers may be solitary, or grouped in umbellate panicles, umbels or corymbs. They are usually greenish yellow to white in colour and have numerous stamens which are bent inwards at first. The four petals are joined together to form a cap, and this is pushed off by the stamens as they unfold. The conical receptacle becomes woody as the fruit is formed, and the loculicidal capsules open by pores arranged in the shape of a cross. The genus contains about 600 species concentrated in Australia. The tallest tree in the world is *E. amygdalina*, which reaches a height of 150m. The whole plant, especially the leaves, is rich in ethereal oils, and the timber is used in many ways.

The Tasmanian Blue Gum, *E. globulus*, was often planted to drain swamps, the home of the malaria-carrying mosquito. As it is a very quick-growing tree, it is often used for afforestation. However, because the leaves contain ethereal oils they decompose only slowly, preventing the formation of undergrowth, and leading to the destruction of the natural ecosystem of the forest.

## Calabash Tree

45

*Crescentia* spp.

**Trumpet-flower family** Bignoniaceae

The genus *Crescentia* has five species native in the New World as far north as Florida. The best known is *C. cujete*, the Calabash Tree, which has entire leaves, broadest near the rounded apex, and is often cultivated. Another widespread species is *C. alata* which has a digitately three-lobed leaf and winged leaf-stalk. Both species reach a height of 8–10m, have a slightly scaly bark, and in their general appearance resemble an apple tree. The flowers are produced directly from the branches and young stems (cauliflory), last only one night and are pollinated by bats. They are broadly tube-shaped, constricted below the middle, and brownish red in colour. The one-celled ovary, formed from two united carpels, develops into a spherical to ellipsoid capsule often nearly as large as a human head and containing numerous flattened pumpkin-like seeds embedded in pulpy tissue. When dried, the shells of the fruits become hard and they can then be used as musical rattles, or if painted and carved as ornaments, or when cut in half as ladles or drinking vessels.

## Madagascar Dragon Tree

46

*Dracaena marginata* Lam.

**Lily family** Liliaceae

The genus *Dracaena* comprises 80 species, distributed from tropical and subtropical Africa and Asia to Australia. The name of the genus is derived from the Greek word 'drakaina' (female dragon). Systematically, the genus *Dracaena* is closely related to the genus *Cordyline*, but may be distinguished from it by the roots which are orange or yellow and often pleasantly scented, and by the number of ovules, solitary in each of the three cells of the ovary in *Dracaena*, many in *Cordyline*. However, the botanical separation of the genera is still not completely resolved and species may be assigned to one genus or the other according to the authority concerned.

The Madagascar Dragon Tree is endemic to that island, and in the open forms a graceful plant several metres high, its thin, branched shoots chequered by the scars of fallen leaves. The slender trunk is topped by a loose tuft of flat, narrowly linear to lanceolate leaves and bases of which partially encircle the stem. These grow up to 50cm long, stand out stiffly at first but bend down as they become old. The midrib is clearly visible on the underside, and the glossy leaf-blade is dark green with a brownish red margin. Although the woody stem of the Madagascar Dragon Tree appears very slender, it nevertheless grows with age, and members of the genus *Dracaena* may be readily distinguished from palms by their proportionally thicker trunks.

## Traveller's Tree 47, 49

*Ravenala madagascariensis* Sonn.

**Banana family** Musaceae

The Traveller's Tree comes from Madagascar and the fan-like arrangement of its leaves renders it quite unmistakeable (Photo 47). At first the young plants have no stem as the leaves develop on an underground rootstock. As the age of the plant increases a trunk 8–10m high is formed, crowned by a 6–8m wide fan of banana-like leaves slit from the margin to the midrib into narrow strips. The boat-shaped bases of the leaf-stalks form two ranks which overlap each other closely (Photo 49) and can hold 1–2 litres of liquid which may be used as drinking water in cases of emergency, hence the English name for this plant.

The inflorescences arise from the leaf-axils, and the numerous flowers are arranged in groups in the axils of the boat-shaped bracts. The flowers have six white perianth-segments in two whorls, the middle one of the inner whorl being a little shorter and narrower than the rest. There are six stamens surrounding the inferior, three-celled ovary. The walls of the ovary exude nectar attracting various kinds of birds which pollinate the flowers and allow the woody capsules to develop. The seeds have a bright blue outer covering which produces a pleasantly scented, antiseptic oil. A sugary sap can be obtained from the trunk.

## Spineless Yucca 48

*Yucca elephantipes* Regel

**Lily family** Liliaceae

The Spineless Yucca is a striking plant on account of its woody stem, broadened at the base, which can reach a height of 8m. Its natural distribution extends from Mexico to Guatemala, and it is often cultivated as a hedge-plant. The upper 40–60cm of the stem bears a number of sword-shaped leaves, 60–100cm long, 5–8cm broad at the middle and narrowed at the base. When mature, the plant produces a compact inflorescence, 60–90cm long and roughly lozenge-shaped in outline, composed of numerous yellowish white, pendulous flowers. The three-celled ovary with its short thick style develops into a capsule containing numerous black seeds. The fleshy flowers can be boiled and used as a salad.

# Cycads

The cycads represent an ancient group of plants that in earlier times, particularly during the Jurassic period, grew on the earth in great variety. The 100 or so members that are still alive today are consequently the last representatives of an order of evolutionary importance, the remains of which are scattered throughout the world. These are living fossil plants, a fact emphasised by their primeval appearance. Nevertheless, a number of them are highly decorative.

Systematically, cycads occupy an important transitional stage between conifers and true flowering-plants. Therefore, from a botanical point of view, they are related neither to palms nor to ferns, and have only acquired the popular name of 'fern palm' from their outward appearance which in the larger forms is reminiscent of a palm tree. The smallest cycads have a turnip-shaped, underground stem with pinnate leaves 8–10cm long, the largest have a stem up to 18m high bearing an imposing crown of palm-like leaves a metre or so in length. In that case the stem is often branched. From an ecological viewpoint, most cycads grow in extremely dry localities, which is reflected in the anatomical structure of their leaves. Some of the very slow-growing species attain an age of 1000 years. Amongst these is the Mexican Fern Palm, *Dioon edule*, from the stem of which a high quality sago is obtained.

All species of cycads are dioecious. In male plants the stamens derived from modified leaves are arranged in large cones at the top of the stem between the frond-like leaves. On the underside of these modified leaves are numerous pollen sacs. The seed-cells or spermatozoids, which develop from the pollen-grains and effect fertilisation, have a fringe of hairs to facilitate movement, and with a diameter of 0.3mm are the largest in the plant and animal kingdoms. In female plants the carpels are not united to form an ovary but show clearly that they also are derived from leaves. They often still possess a stalk and a pinnate blade on which there are large, separate ovules visible to the naked eye. The carpels may also be united in cones or may be found as leaf-like organs, densely covered with golden brown hairs, arranged alternately with the foliage leaves at the end of the stem. The egg-cells which arise within the ovules can be as long as 6mm!

## Fern Palm

**50, 51**

*Cycas circinalis* L.

**Fern Palm family** Cycadaceae

The Fern Palm is cultivated everywhere in the tropics because of its ornamental growth. It is native in Ceylon, southern India and in the Indomalayan Archipelago. It is a plant with a stem 2–3m or even up to 5m in height, the surface of which is covered by the brown bases from which leaves have fallen. The crown consists of very decorative pinnate leaves which arch over slightly. They are 2–3m long and have 50–60 narrowly lanceolate leaflets on each side of the midrib. When young, the leaves are coiled inwards like fern fronds. In female plants the golden brown carpels form dense cones at the top of the stem (Photo 51). In the course of further growth they are pushed aside by the developing leaves which are arranged alternately with them. On the margin, above the stalk-like portion, they bear 5 or 6 large, pale olive green ovules. In male plants the stamens are united into an impressive cone 30–50cm long (Photo 50). A single cone produces millions of pollen-grains which are carried by the wind to the ovules of the female plants.

## Bread Tree 52

*Encephalartos altensteinii* Lehm.

**Fern Palm family** Cycadaceae

Many of the cycads of South Africa have only a very local distribution, but the Bread Tree is occasionally planted in gardens. It grows up to 5m high and forms a crown of pinnate leaves 1–2m in length. The numerous shining leaflets have 3–5 sharp teeth on the edge, and these are modified into thorns at the base of the leaf. The female plants produce large cones which may be 50–60cm long and 30cm in diameter. Each carpel bears on the underside of the scale-like end portion two ovules which develop into hard-shelled seeds rich in starch. At maturity the cones weigh 20–30kg, and in the case of the closely related species *E. caffer* as much as 42kg. The name of the genus, *Encephalartos*, means 'bread in the head' and was given because the plants contain starch both in their seeds and, especially, in the stem. In earlier days *E. transvenosus* had special significance in the Transvaal and all the plants were placed under the protection of the Bantu rain-queens.

# Palms

## Cuban Royal Palm 53, 54

*Roystonea regia* (H.B.K.) O.F. Cook

**Palm family** Palmae

The Cuban Royal Palm is one of the most beautiful and popular ornamental palms of the tropics. Its smooth, light grey trunk, slightly barrel-shaped when young, reaches a height of 25m and is crowned by huge, pinnate leaves. Its majestic appearance makes it particularly suitable as an avenue tree, though it is somewhat demanding in soil conditions. The home of this species is the West Indian islands, especially Cuba, where it thrives best on a deep, nutrient-rich soil. Its leaves reach a length of up to 8m, of which 2m is the smooth green leaf-base which, being soft and pliable, is used for all kinds of purposes, e.g. as a covering for huts and as packing for Cuban tobacco. As with many other palms, the leaf-blades themselves serve as a roof-covering.

The inflorescences arise below the leaves, at first forming spadices 1.0–1.5m in length and directed upwards at an angle. After the handsome spathe surrounding the spadix has opened, the inflorescence, consisting of numerous pale yellow flowers, is able to unfold. One can usually find spadices and inflorescences in various stages of development on the same tree. In female flowers, the three-carpelled ovary develops into a fruit which is green at first, later red, and finally dark brown to blue-black. The fruits are rich in oil and carbohydrates and are used as food for pigs. A single cluster of fruit can weigh as much as 50kg.

The trunks of Cuban Royal Palms often exhibit slight swellings and constrictions, the result of the changing nutritive and climatic conditions. Like all other palms the trunk develops equally to its final diameter. The closely related South American Royal Palm, *R. oleracea*, reaches a height of 40–50m. Its growth is perhaps more impressive but less well balanced when compared with that of the Cuban Royal Palm. The growing point of the South American Royal Palm provides an excellent palm cabbage.

## Spindle Palm

55

*Hyophorbe verschaffeltii* Wendl.

**Palm family** Palmae

The generic name *Hyophorbe* is derived from the Greek *hys*, *hyos* (pig) and *phorbe* (food) and refers to the fact that the fruits of this palm, like those of many other palms, have been used to feed pigs. The genus comprises a few species native in the Mascarene Islands. Some of these are characterised by an unbranched trunk which is swollen half-way up or just below the crown of leaves. This is the case in the species illustrated which comes from the island of Mauritius. Its pinnate leaves are 2–3m long and very broad, but they are delicately curved. The leaflets are very long and are shaped like a sword. The leaf-stalk is often somewhat reddish, and usually has a golden yellow stripe on the underside and a brown leaf-sheath at its base.

## Sealing-wax Palm

56

*Cyrtostachys renda* Bl.

**Palm family** Palmae

The Sealing-wax Palm is one of the most beautiful and ornamental palms in tropical gardens. Its home is the island of Sumatra. The genus contains 11 species, extending along the Indomalayan Archipelago to the Solomon Islands, but in S.E. Asia it is the Sealing-wax Palm which is most frequently cultivated as a decorative plant. It is a graceful feather-palm with a slender smooth trunk up to 10m high, topped by a dense tuft of bluish green pinnate leaves which have a glossy bright red leaf-sheath and a red midrib. The leaflets are stiffly directed forwards. The inflorescences appear below the fronds, and in the spadix state are enclosed in two fully formed bracts. The flowers themselves are arranged in groups of three, the middle one in each group being female, and developing later into the fruit botanically regarded as a berry.

## Miniature Date Palm, Pygmy Date Palm

57

*Phoenix roebelenii* O'Brien

**Palm family** Palmae

Like the Canary Date Palm and the true Date Palm this species is also dioecious. The Miniature Date Palm is native in Assam and Cochin-china, whereas most of the 13 species of *Phoenix* come from Africa. It is a relatively small but very decorative tree, its trunk only reaching a height of 2m. *P. roebelenii* can be distinguished from closely related species by its young shoots which are covered with a fine powder. The three inner perianth-segments of the male flowers are narrow and flap-like, while all six perianth-segments of the female flowers are broader and overlap at the edges. The Miniature Date Palm is fairly hardy and is a favourite house-plant in temperate latitudes.

## Alexandra Palm 58

*Archontophoenix alexandrae* (F. Muell.) Wendl. & Dr.

**Palm family** Palmae

This is a small, quick-growing feather-palm which likes full sun. Its slender trunk grows up to 8m high, and bears a crown of pinnate leaves which bend over in graceful curves. These relatively stiff, but very beautiful leaves can hang down a long way, and, together with the coral-red fruits, give the plant a particular charm. The much-branched, pale green to cream-coloured inflorescences appear at the top of the grey trunk just below the point where the leaves with their long sheaths are produced. As often happens in palms, the male and female flowers are separated. After fertilisation, the ovaries of the female flowers develop into coral-red fruits, over 2cm long, elliptic in cross-section and with a distinct point (Photo 59). The plant is native in Australia. The genus comprises four species which are distributed from tropical Asia to Queensland.

## Manila Palm, Christmas Palm 59

*Veitchia merrillii* (Becc.) H.E. Moore

**Palm family** Palmae

The genus *Veitchia* is named after James Veitch, the owner of a famous nursery in Chelsea in the 19th century, but until recently the species was placed in the genus *Adonidia*. Its home is in the Philippines, New Hebrides, New Caledonia and the Fiji Islands, but it is cultivated nowadays as a decorative plant throughout the tropics. It is a palm of medium height with a trunk up to 5m and a highly ornamental crown of pinnate leaves. The leaves, which arch over to a length of up to 2m, have numerous sword-shaped leaflets up to 90cm long. This palm is monoecious, the male and female flowers appearing in inflorescences which arise, stiffly erect at first, from the top of the trunk, just below the smooth green leaf-bases. The three-celled ovaries develop into shining red fruits which are characterised by a striking cup-shaped depression at the end and enhance the ornamental aspect of the plant. In Florida, five other species of the genus are in cultivation.

## Chinese Fan Palm 60

*Livistona chinensis* (Jacq.) R. Brown ex Mart.

**Palm family** Palmae

The genus *Livistona* was named in honour of a Scottish nobleman, Patrick Murray, the Laird of Livingston in West Lothian, and comprises 30 species which are distributed in tropical Asia and from the Malayan archipelago through New Guinea to eastern Australia. The species illustrated occurs in southern China and is widespread as an ornamental tree. It reaches a height of more than 10m and the crown is 7–8m across. Its shining fan-shaped leaves are almost circular in outline, and are composed of many segments, those at the centre being somewhat longer than the others. The leaves arch over and the long fibres hanging from the ends of the segments add to the decorative effect. There are usually backwards directed spines along the lower half of the leaf-stalk. This palm can easily be confused with *Washingtonia filifera* which is found in California and northern Mexico.

## Queen Palm

61

*Arecastrum romanzoffianum* (Cham.) Becc.

**Palm family** Palmae

The distribution of the Queen Palm extends from central Brazil to Paraguay, Uruguay and north Argentina. In south Brazil it stretches eastwards towards the Atlantic coast as far as the dune regions. It is a quick-growing tree, reaching a height of up to 12m. The solid, column-like trunk bears a broad crown consisting of numerous, curved, overhanging pinnate leaves, which are up to 6m long. The very narrow leaflets, up to a metre in length, are arranged either singly or in tufts along the midrib of each frond. The palm will also thrive in subtropical climates.

## Petticoat Palm

62

*Copernicia macroglossa* Wendl. ex Becc.
Syn.: *C. torreana* Léon

**Palm family** Palmae

The genus *Copernicia* is named after the astronomer Nicolaus Copernicus and comprises 44 species from the Antilles to tropical South America. There are 25 species on Cuba alone, including the Petticoat Palm. This is only 7–8m high, in contrast to related species which usually grow quite tall. The upper part of the relatively short stem remains clothed by a dense covering of leaves which hang on for many years after they are dead and form a skirt round the trunk, hence the English name for the plant. The circular, fan-like leaves are made up of numerous leaflets radiating from the end of the short stalk. Of all the species in this genus, the Carnauba Palm, *C. cerifera*, which is found in Brazil, Paraguay and Argentina, is the most important source of wax. Carnauba wax is a constituent of high-quality polishes, and is also used in the manufacture of paints and varnishes, candles, carbon paper, films, records, lipsticks, ointments and many other products.

## Ruffled Fan Palm

63

*Licuala grandis* Wendl.

**Palm family** Palmae

The genus *Licuala* comprises more than 80 species, distributed throughout tropical east Asia, the Malayan Archipelago, New Guinea and northern Australia. One of the most striking species is *L. grandis*, the Ruffled Fan Palm, which comes from the island of New Britain north of New Guinea. It reaches a height of 2m and bears numerous, almost circular leaves on a short trunk. The dark green leaf-blades are like round fans and are slightly lobed along the margin, each lobe being divided at its tip into two teeth by a small incision. The slender trunks of *Licuala* palms are often tube-like in appearance and thrive best in a damp climate.

## Sentry Palm

64, 65

*Howea forsteriana* (F. Muell.) Becc.

**Palm family** Palmae

Sentry Palms originate from the Lord Howe Islands in the Pacific Ocean east of Australia. Two species are known, both of which are popular ornamental palms also grown outside the tropics in pots as house-plants. Sentry Palms can reach a height of 8–10m and have a smooth, permanently green trunk, the surface of which is patterned only by the scars of fallen leaves. The handsome pinnate leaves are 2–3m long, and in the species *H. forsteriana* are bright blue-green in colour, and stand out almost horizontally from the top of the trunk (Photo 64). The branched inflorescences later bear ellipsoid fruits. In contrast to the previous species, the inflorescences of *H. belmoreana* are unbranched and remain enclosed in the protective bract for a long time. The fruits of this species are almost globular in shape. In its natural habitat *H. forsteriana* grows taller than its companion species, and lacks the thickening at the base of the trunk which is characteristic of *H. belmoreana*.

## Wine Palm, Sago Palm

66

*Caryota urens* L.

**Palm family** Palmae

The genus *Caryota* has foliage which is quite different from all other palm genera. *C. urens*, the Wine Palm, reaches a height of 10m and is native in India, Ceylon and the Malay Peninsula. It is both fast-growing and tough. The smooth grey trunk is marked with rings where the older leaves have fallen off. The leaves arch over slightly, and are doubly pinnate in form, the individual leaflets being curiously lozenge- or wedge-shaped, hence the English name for the genus, Fish-tail Palms. The name of the species was probably suggested by the burning sensation experienced when the fruit is eaten. The inflorescences hang down in a very decorative way, and consist of densely crowded male and female flowers which open in turn from the top to the base. The male flowers develop slightly before the female. An excellent sago is obtained from the pith of the trunk. The juice is tapped at the beginning of flowering time, and is used in India as a drink. When fermented it becomes palm wine or toddy. The leaf-sheaths are the source of Kittul fibre which is much used for making brushes, brooms, ropes and also baskets. The closely related Clustered Fish-tail Palm, *C. mitis*, is distinguished by its more open foliage and mild-tasting fruit. In all, 27 species of *Caryota* are recognised originating in tropical Asia and Australia.

## Canary Date Palm

67, 68, 69

*Phoenix canariensis* hort. ex Chabaud

**Palm family** Palmae

The Canary Date Palm is closely related to the true Date Palm, but differs in having a shorter trunk and a denser crown of pinnate leaves. Both plants belong to a genus which comprises 14 species distributed from the Canary Islands across North Africa to India, Sumatra and Formosa. The individual species are not easily distinguished from each other. Because it originates in the Canary Islands, the Canary Date Palm is also found in the cooler upland regions of the tropics. It is cultivated even more often in the subtropics where it is a widespread and favourite ornamental palm. When mature it is a majestic tree up to 15m high with a trunk marked with the scars which show where fallen leaves had grown and crowned by an imposing mass of pinnate leaves 5–6m in length (Photo 67). In many cases the lower leaflets of the fronds are reduced almost to thorns. As the plant is dioecious, trees are either male or female. The male inflorescences are more strongly branched (Photo 68). The bract or spathe, which encloses the inflorescence before it unfolds, remains for a long time. The female inflorescences (Photo 69) are less striking and are composed of smaller, rather insignificant flowers which develop into slightly fleshy fruits. In the tropics the little cavities above the leaf-bases are frequently occupied by epiphytic ferns. One very well known decorative palm in the genus *Phoenix* is *P. roebelenii* which comes from India and has delicate pinnate leaves up to 1m long. In temperate zones it is often grown in pots as a house-plant.

# Shrubs

## Poinsettia, Christmas Star

70

*Euphorbia pulcherrima* Willd.

**Spurge family** Euphorbiaceae

As the Poinsettia blooms in the winter months, it is also known as the Christmas Star. It has acquired this name from the rosette-like arrangement of the bright red bracts at the end of the stems. The shrub is native in Mexico and Central America and it reaches a height of 3–4m. The simple or branched, brown stems bear long-stalked leaves, 7–15cm in length and variable in shape. The leaf is wedge-shaped at the base and usually long-pointed at the apex, but may be ovate-oblong, entire, fiddle-shaped or lobed. At the top of the stems are the strikingly blood-red, usually lanceolate, bracts which may also be pink, cream or pale green in the many cultivated forms. The flowers themselves are insignificant and are arranged above the bracts in small umbel-like groups. In the centre stands a single female flower surrounded by numerous male flowers which are reduced to stamens only. Each group of male and female flowers is surrounded by an involucre of tiny bracts. The ovary of the female flower develops into a three-celled capsule.

The Poinsettia, called 'flor de pascuas' in Spanish, flowers for several months during the winter, and is a favourite decorative plant in all tropical gardens. It is one of the 'short-day' plants, and its time of flowering is connected with the shorter periods of daylight occurring at this season of the year. In its natural habitat the Poinsettia loses its leaves during the dry season. The shrub flowers better if the stems are pruned after flowering.

## Chinese Hibiscus

72

*Hibiscus rosa-sinensis* L.

**Mallow family** Malvaceae

The Chinese Hibiscus is one of the best known and most splendid ornamental shrubs in the tropics. Its native country is thought to be China, from where it has spread to tropical gardens everywhere. Since the earliest times it has been used as a decorative plant because it flowers continuously from spring to autumn.

It is a shrub with slightly stiff, spreading branches, 2–5m high. The leaves are ovate in shape with a long tip and a toothed margin. The coral-red corolla is 10–15cm across and makes an attractive contrast with the shining, dark green calyx formed from five united sepals. In the centre of the corolla are numerous stamens, united into a tubular column which protrudes beyond the petals. At its upper end are numerous spreading anthers, with the five stigmas radiating from the top of the style and forming a crown above them. The flowers only last one day.

The Chinese Hibiscus is cultivated in numerous forms, including double ones, in all tropical regions where it replaces the rose, but lacks its scent. Humming-birds and other nectar-feeding birds are often attracted by the flowers. The fruit consists of a five-celled capsule. Juice from the petals of Hibiscus flowers was formerly used to blacken the hair and eyebrows and also served as shoe-blacking.

## Fringed Hibiscus 71

*Hibiscus schizopetalus* (Mast.) Hook.f.

**Mallow family** Malvaceae

The Fringed Hibiscus is cultivated almost as widely as the Chinese Hibiscus, but the home of this species is in East Africa. In growth it resembles the Chinese Hibiscus but its foliage is less dense. Its slender, usually arching stems bear stalked, ovate-elliptic leaves with a toothed edge. Like the Chinese Hibiscus there are variegated-leaved forms. The two species have been used as parents for a large number of hybrids.

The pale red flowers have delicately fringed petals and produce a striking effect as they hang down from the twigs on long, jointed stalks. The bases of the incised and reflexed petals are enclosed in a tubular calyx. The tubular column formed by the united stamens bears a group of spreading anthers near its tip and protrudes from the turban-like corolla even further than that of the Chinese Hibiscus. The united styles project from the hollow staminal column, their spreading stigmas giving a final touch to the bizarre character of this flower. The Fringed Hibiscus resembles the Chinese Hibiscus in flowering most of the year and consequently it is a splendid ornament in almost all gardens in the tropics.

## Wax Mallow, Turk's Cap, Cardinal's Hat 73, 74

*Malvaviscus arboreus* Cav.

**Mallow family** Malvaceae

As a rule the Wax Mallow is a small shrub, but it can become a small tree. Its generic name means 'sticky mallow' which it acquired from the abundant mucilage present in the species. The native distribution of the plant extends from Mexico, Central America and the West Indies to Peru and Brazil. The long-stalked, ovate-lanceolate, pointed leaves have toothed or entire margins and are usually densely hairy on the underside. The flowers are up to 5cm long and arise from the axils of the leaves. Like many of the Malvaceae they have an epicalyx consisting of numerous tiny leaves pointing upwards. Next come the five green sepals and then the five considerably larger petals, fiery red, and twisted in the bud stage (Photo 73). The staminal column, composed of numerous stamens, protrudes from the corolla, and the five spreading style-branches each with a stigma at the end, extend even further. When the hummingbirds and other nectar-feeders visit the flowers to obtain the abundant nectar which collects at the base of the calyx, they dust the stigmas with pollen that has adhered to their feathers. The fruit of the Wax Mallow is a round berry that breaks up into five parts. Besides the typical form with bright red flowers there are also cultivated forms with flowers of a delicate pink colour (Photo 74) or even yellow.

## Croton

**75**

*Codiaeum variegatum* L.

**Spurge family** Euphorbiaceae

The Croton is extremely variable in form and colouring. It is native in Indonesia and New Guinea. Numerous forms arise by hybridisation in the South Sea area. The shrub grows to a height of 2.5m, and has elliptic, lanceolate, or linear leaves, which may be lobed or in extreme cases reduced to the midrib only. As the colour can range from pure green to yellow and different shades of red or be patterned with yellowish green or orange, the variability of the leaves is further increased.

The long inflorescences are composed of rather insignificant unisexual flowers. The male flowers usually bend over slightly and have 5 or 6 small petals which overlap in bud and have glands between them. The 15–30 long stamens are quite conspicuous. The female flowers are usually entirely lacking in petals. Like many Euphorbiaceae they possess a three-celled ovary with three spreading stigmas. The ovary ripens into a capsule which eventually splits into three parts. The female flowers stand upright but are less striking than the male. The Croton (not to be confused with the botanical genus *Croton*) is found in all tropical gardens. The oil obtained from the seeds has a strongly purgative effect.

## Lantana

**76**

*Lantana camara* L.

**Verbena family** Verbenaceae

The natural distribution of the Lantana extends from the tropical regions of South America northwards to Texas and South Carolina. From there it has spread as an ornamental plant to all parts of the tropics. It has become naturalised in many places and forms impenetrable thickets in Ceylon and Java.

It is an often prickly shrub which can reach a height of 1.5m or may creep about on the surface of the ground. Its square stems bear pairs of ovate to heart-shaped leaves, pointed at the apex and covered with a network of veins. The leaves are toothed along the margin and are hairy, especially on the veins. The flowers have a five-lobed corolla and a small calyx and are arranged in compact, rounded spikes. The four stamens are attached at the middle of the corolla-tube. The colour of the flowers is very variable and can change in the course of the flowering period. There are genetically pure forms (clones) where the colour changes from orange through bright yellow to crimson, others go from pink to a fiery red. There are also forms which are violet, lilac or white in colour, or in intermediate shades. The two-carpelled ovary develops into a blue-black, berry-like fruit which is highly decorative but dangerously poisonous. The aromatic leaves are used for medicinal purposes and also to make a stimulating tea. Lantana is becoming more popular in temperate regions where it is grown as an annual for summer bedding.

## Red Flag Bush 77

*Mussaenda erythrophylla* Schum. & Thonn.

**Madder family** Rubiaceae

The genus *Mussaenda* contains more than 100 species distributed throughout the tropics of the Old World. It includes shrubby, climbing and epiphytic plants. The Red Flag Bush is a striking decorative shrub which comes from tropical Africa, and has broadly elliptic to heart-shaped, pointed leaves set in pairs along the stems. The inflorescence with its bracts and stalks of an intense red is especially typical. The flower has five red sepals, one of which is enlarged, strongly lobed, and directed outwards. The corolla consists of five united petals, which are white in colour and contrast with the bright red of the other parts of the inflorescence. It is interesting to see how plants on widely separated continents have developed in a similar way. For example, another member of the Rubiaceae, *Warscewiczia coccinea* from the tropical rain-forest of Peru, has a similarly enlarged, red sepal. To some extent both of these species resemble the Poinsettia, but in the case of that plant it is not the sepals but the bracts, coloured by the red pigment anthocyanin, which catch the eye.

## Chenille Plant 78

*Acalypha hispida* Burm.f.

**Spurge family** Euphorbiaceae

The genus *Acalypha* has leaves that resemble those of the Nettle, although it is not related botanically to that plant. The Chenille Plant is thought to have been native in Indonesia. It develops into a shrub several metres high with conspicuously long inflorescences that hang down from the axils of the leaves. These bright red, catkin-like inflorescences are composed of numerous female flowers with feathery styles but an inconspicuous perianth. The male flowers also arise from the leaf-axils but, as the species is dioecious, they are found on separate plants. When mature, female plants produce three-celled fruits. The leaves have a distinct stalk and are arranged alternately on the stems. There are 430 species in the genus, distributed throughout the tropics and subtropics, and they include herbaceous plants and trees as well as shrubs. The Chenille Plant is often found in and around villages where it is used as a hedge-plant.

## Copperleaf 79

*Acalypha wilkesiana* Muell.Arg. 'Macrophylla'

**Spurge family** Euphorbiaceae

The genus *Acalypha* contains numerous species and ornamental cultivars and includes *A. wilkesiana*, a plant with a particularly wide range of colour forms. This species becomes 2–3m high and originates in the South Sea Islands. Its leaves are broadly ovate to heart-shaped and are pointed at the end. They are almost always variegated and their colour ranges from red and carmine pink through bronzy green to dark and pale green. The various colour-patterns are caused by the individual layers of cells of the leaf having different pigments in varying combinations. In the places where the leaf is pure green, the colour of the chloroplasts has not been modified by other pigments. The purely red areas exhibit no green chlorophyll granules, although the outer layers of the leaf contain the soluble red pigment anthocyanin in abundance. By a combination of these two conditions mixed colours arise in great variety. The inflorescences of this plant are inconspicuous, being much shorter than the leaves and almost hidden. In a few cases cultivar names have been given to distinguish certain colour-forms whose origin can no longer be ascertained.

## Flame of the Woods **80**

*Ixora coccinea* L.

**Madder family** Rubiaceae

The genus *Ixora* gets its name from a Malabar deity and comprises 400 species native in tropical Africa and India. The species *I. coccinea* is considered to be one of the loveliest decorative shrubs of the tropics because of its scarlet inflorescences. Its shortly stalked leaves, lanceolate to ovate in shape and pointed at the tip are arranged in pairs on the stems. It flowers when it is still a small shrub and may eventually reach a height of 5m. The fiery red flowers have four, rarely five, corolla-lobes which spread out at right-angles to the slender corolla-tube. The flowers are arranged in rounded umbels with individual flowers, still in the bud stage, poking out like little arrows above the rest of the cluster. The two-celled ovary, with two short styles develops into a berry. The species concerned, which comes from India, has been crossed with other species, so that hybrids are often found in cultivation with flowers of a salmon-pink colour or other paler shades. The name 'Flame of the Woods' was given to the typical plant because of the way its brilliant red flowers, which remain open for a long time, contrast with the glossy, dark green leaves.

## Pagoda Flower **81**

*Clerodendrum paniculatum* L.

**Verbena family** Verbenaceae

The genus *Clerodendrum* includes both erect and climbing species. One of the most familiar decorative plants in the genus is the Pagoda Flower which is native in south-eastern Asia from Burma, through Malaya to China. It is conspicuous for its paniculate inflorescences and its long-stalked, heart-shaped to five-lobed leaves with prominent veins. The striking panicles of flowers stand out from the luxuriant foliage. The flowers have a bell-shaped, five-lobed calyx and a very long and narrow corolla-tube which opens out into five wide-spreading lobes. The four stamens and style project some way beyond the corolla-tube. The ovary develops into a berry-like drupe.

## Glory Bush **82**

*Tibouchina semidecandra* (Schrank & Mart.) Cogn.

**Melastoma family** Melastomataceae

The genus *Tibouchina* comprises 250 species in southern Brazil and the Andes. The south Brazilian species *T. semidecandra*, the Glory Bush, is one of the most ornamental, flowering during the winter from November to May and sometimes at other times of the year as well. It is a beautiful shrub up to 6m high with pairs of leaves more than 12cm long and drawn out to a pointed tip. The highly decorative, purplish pink flowers are more than 10cm across and in the bud stage are enclosed in two bracts which fall off as the flower opens. The stamens have a rather strange appearance with bent filaments, long-shaped anthers and an appendage on the inner side. In addition to the pink form, violet and purple forms are often found.

## Barbados Pride, Dwarf Poinciana **83**

*Caesalpinia pulcherrima* (L.) Sw.

**Senna family** Caesalpiniaceae

Barbados Pride or Dwarf Poinciana is related to the Flamboyant, and the generic name commemorates the Italian philosopher, doctor and botanist Caesalpinus. It is one of the loveliest decorative shrubs in tropical gardens and originally came from the islands of the West Indies. It can reach a height of 6m and its doubly pinnate leaves have a wonderfully delicate appearance. Each leaf has a midrib with 3–9 lateral ribs bearing 6–12 pairs of ovate-linear leaflets clipped rather abruptly at the end. Above the ornamental leaves are the clusters of flowers, usually red in colour but quite often orange or yellow also. They are conspicuous on account of their protruding stamens and single long style, developing later into flattened, pendulous pods, up to 12cm long and 3cm broad. When these become ripe they turn brown and may be slightly twisted. The plant was placed earlier in the genus *Poinciana*, and like *Delonix regia*, also known as 'Flamboyant' in the tropics. The flowers and leaves are supposed to be able to reduce fever, and in addition the leaves have purgative properties.

## Coral Plant, Fountain Bush **84**

*Russelia equisetiformis* Cham. & Schlecht.

**Figwort family** Scrophulariaceae

The genus *Russelia* is named after the English naturalist Alexander Russel, and comprises some 20 species native in tropical America. The Coral Plant from Mexico is the species most frequently cultivated. It grows to a height of 80–120cm and has slender, green stems, often branched and ending in a number of fine, slightly overhanging twigs. The plant has lanceolate to linear leaves only at the base of the stems. On the upper parts the leaves are reduced to scale-like teeth. The scarlet flowers are arranged in loose racemes. Each flower has a small calyx and a tubular corolla 2–5cm long, the five lobes forming two lips at its mouth. Enclosed within the corolla-tube are four stamens. Like many yellow and red-flowered plants in the tropics, *Russelia* is visited by humming-birds and other nectar-feeders.

## Yellow Oleander **85**

*Thevetia peruviana* (Pers.) Schum.

**Dogbane family** Apocynaceae

The Yellow Oleander is closely related to the Oleander and grows as a stately bush or small tree in the tropics of Central and South America. Its branches bear narrowly linear to lanceolate leaves, shiny on the upper side, which are reminiscent of the Oleander. At the ends of the stems are the wax-like flowers, twisted in bud, which have a long corolla-tube. Because of their colour the plant is known as the 'Yellow Oleander'. The two-carpelled ovary develops into a hard, broadly triangular fruit containing two large, flattened seeds, which are often worn as ornaments or carried in the pocket to bring good luck. The plant contains in its milky juice the glycoside thevetin, and all parts of the plant are dangerously poisonous. It is cultivated mainly in Hawaii. From there the fruits are exported so that a heart drug can be obtained from them. Another form, with orange-coloured flowers, is in cultivation. Amongst the Hindus the Yellow Oleander is often chosen as an offering to the god Siva.

## Powder-puff

86

*Calliandra* spp.

**Mimosa family** Mimosaceae

The genus *Calliandra* comprises about 150 shrubs and small trees which originate almost exclusively in tropical and subtropical America. Their lovely heads of flowers make them a favourite among decorative plants. The broadly brush-shaped inflorescence consists of a mass of individual flowers with numerous silky, shiny stamens. The colour of the stamens usually varies from bright red to deep dark red, but may be pink or even white according to the species concerned. The flowers have four petals, and a one-celled superior ovary which develops into a usually flattened pod with thickened edges. When ripe, the pod springs open and, because of the thickened margins, the two halves curl round into spirals. The species most often encountered are *C. fulgens* and *C. haematocephala*, both with red flowers.

## Candle Bush

87

*Cassia didymobotrya* Fresen

**Senna family** Caesalpiniaceae

The genus *Cassia* is predominately native in tropical and subtropical America. Amongst the shrubby species the Candle Bush is the one most frequently cultivated. Its natural distribution is in Africa though it is in cultivation all over the tropics. The plant grows to a height of 2–3m and bears handsome pinnate leaves with broadly elliptic leaflets which are not quite at right angles to the midrib. The bright yellow flowers are arranged in dense clusters at the ends of the erect stems giving the effect of a candelabra with numerous candles. The ovaries develop into winged pods with many seeds.

## Red Bauhinia

88

*Bauhinia galpinii* N.E. Brown

**Senna family** Caesalpiniaceae

The Bauhinias include small trees, shrubs and lianes, and the tree-like and shrubby species have flowers which vary considerably in size and colour. *B. variegata*, the Orchid Tree, has strange, pink-coloured flowers reminiscent of an orchid flower, and the basic structure, characteristic of the Caesalpiniaceae, is only apparent on closer inspection. The fruits of the Orchid Tree are smooth, brown pods. In addition to pink-flowered species there are some with vermilion flowers such as *B. monandra* from the Antilles. The leaves of this plant are up to 15cm long, split into two parts, and distinctly net-veined. The single fertile stamen reaches a length of 4cm. The pods are linear, flat and up to 20cm long. Another species often cultivated is *B. picta*. A particularly striking tree-like species from Brazil is *B. alata*, whose dark purple flowers may be as much as 18cm across.

Lianes are found amongst the Bauhinias in all tropical regions of the world, and these climbing species attach themselves to the supporting trees partly by means of spirally coiled tendrils. Other species develop flattened stems which bend backwards and forwards as they climb producing a snake-like or even step-like effect. In fact, these step-like stems are commonly known as 'monkey staircases'.

## Petrea 89

*Petrea* spp.

**Verbena family** Verbenaceae

The genus *Petrea* was named after the famous English plant-lover Lord Petre, and comprises 30 or so species from Mexico to Panama and the West Indies. They include *P. arborea*, often planted as an ornamental shrub, and *P. volubilis*, the Queen's Wreath, another shrub which climbs to a height of 10m. The latter species has pairs of shortly stalked, elliptic leaves, rough on both sides, and in March and April produces arching racemes of beautiful blue flowers. A white-flowered form of *P. volubilis* is sometimes found. The flowers have five especially long calyx-lobes which are clearly visible before the flowers are fully open and can easily be mistaken for the corolla. The five corolla-lobes themselves are comparatively short and form a small, spreading margin to the corolla-tube. Later on the calyx-lobes turn green, but remain attached to the fruit as wings and aid its dispersal.

## Bottle-brush 90

*Callistemon* spp.

**Myrtle family** Myrtaceae

The members of the genus *Callistemon* owe their beauty to the numerous flowers which are arranged round the stem like a bottle-brush, hence the English name for these plants. Their showy appearance is due to the long, usually bright red or yellow stamens which extend outwards far beyond the tiny corolla. The round, woody fruits, looking like buttons, remain on the stem for a long time. It is interesting to note that the inflorescence is crowned by a group of leaves, for the shoot continues to grow after the flowers have formed. The genus *Callistemon* comprises 12 species native in Australia and Tasmania. The species *C. lanceolatus* is also found as a favourite ornamental shrub in the mediterranean region.

## Guatemalan Rhubarb, Gouty Foot 91

*Jatropha podagrica* Hooker

**Spurge family** Euphorbiaceae

The plant acquired its better known English name from the shield-shaped leaves, three to five-lobed and up to 20cm long, and because it is native in Guatemala. Its natural distribution extends to the West Indies where it forms a shrub up to 60cm high with a conspicuously thick and irregularly swollen stem. It loses its leaves during the dry season. The plant has a branched inflorescence, composed of orange-red, unisexual flowers. The male flowers have a five-lobed calyx, and ten stamens arranged in two whorls. The female flowers have a three-celled ovary with the style divided into several stigmas. The ovary later develops into a roundish capsule. Like all members of the Euphorbiaceae this species has a milky juice. As a decorative plant, it has spread all over the tropics.

## Peregrina — 92

*Jatropha integerrima* Jacq.
*Syn: J. acuminata* Desv.

**Spurge family** Euphorbiaceae

The Peregrina is very variable in leaf-shape and consequently has been given a number of different scientific names. Its long-stalked leaves may be heart-shaped with a distinct point or lanceolate to linear in shape, and either entire or lobed. The shrub grows up to 2m high and has panicles of flowers with small, green calyces and beautiful crimson red corollas up to 2.5cm across. The plant is native in Cuba but is grown as an ornamental plant in all the tropical regions of the New World. The genus *Jatropha* comprises some 150 species, mainly in tropical America but some also in Africa. One important species is the Physic Nut, *J. curcas*, a shrub with ivy-like leaves and mottled seeds, rich in oil, that are the source of a purgative. The remains of the pressed seeds are poisonous, like those of *Ricinus*, and eating only a few seeds may have fatal consequences. *J. multifida* is another species often cultivated for decorative purposes, mainly because of its large, digitate leaves.

## Pink Ball Tree — 93

*Dombeya wallichii* (Lindl.) Benth. & Hook.

**Cocoa family** Sterculiaceae

The Pink Ball Tree is one of the most attractive ornamental plants of the tropics on account of its pink, pendulous, hemispherical inflorescences. Its native country is Madagascar. The form of the inflorescence resembles that of the *Hydrangea* but systematically it is related to the Cocoa plant. It is a small tree or often a spreading shrub 6–8m in height. The leaves are 20–25cm long and broad, softly hairy, and pointed at the tip, rather like a large Lime leaf. The lobes of the heart-shaped base of the leaf overlap at the edges. The large bracts are also heart-shaped with a long, pointed tip and lie closely against the hairy, slightly angular stems which are green in the young state. The compact inflorescences are 12–15cm in diameter and are made up of numerous five-petalled flowers, each one 2cm across with five stigmas spreading out wide above the ovary. After the flowers have faded, the brown inflorescences remain hanging there for a long time. The genus comprises 200 species, mainly distributed in Africa and Madagascar.

## Plumbago, Cape Leadwort — 94

*Plumbago capensis* Thunb.

**Plumbago family** Plumbaginaceae

The genus *Plumbago* contains 20 species which are native in the tropics as well as the subtropics. The most popular species in both these areas is *P. capensis*, from the Cape Province of South Africa, and it is widely cultivated as a decorative plant. The names by which this species is generally known refer to the lead-coloured roots of the plant. It is a small shrub, often used for hedges, with slender, green, more or less erect stems which scramble rather than climb. The leaves are entire, bluntly pointed, and have small scales on the underside. The flowers are pale blue and are arranged in small groups at the ends of the stems. In any cluster the flowers face in more or less the same direction. They have a tubular, five-ribbed calyx and a long, narrow corolla-tube which opens out into five widely spreading lobes. The fruit is a capsule which splits into five parts. Apart from the typical pale blue form there is a cultivated form with white flowers.

### Brunfelsia 95

*Brunfelsia pauciflora* (Cham. & Schlecht.) Benth.
Syn: *B. calycina* (Hook.) Benth.

**Nightshade family** Solanaceae

The genus *Brunfelsia*, popularly known as Yesterday-today-and-tomorrow, was named after the German doctor, botanist and theologian Brunfels. It comprises 30 species distributed throughout Central and South America and the Antilles. *B. pauciflora* is a species which branches from the base into a broadly spreading shrub. The leaves are elliptic to obovate in shape and are pointed at the apex. They are 8–10cm long, glossy above and pale green on the underside. The purple flowers are in small but dense clusters. They are shortly stalked and the corolla-lobes spread out to a diameter of 5cm. The five pale green sepals are united to form an inflated calyx, and the four stamens are attached to the inside of the corolla-tube. The ovary, formed from two united carpels, develops into a leathery capsule with relatively large seeds. This species is native in Brazil as is *B. hopeana*, the source of Manaca root. This contains the very poisonous alkaloid manacine which is used to treat snake-bites.

# Twining and Climbing Plants

Decorative twining and climbing plants are far more abundant in the tropics than in temperate zones, and they include woody plants as well as herbaceous plants. The larger number of tropical species is due to the damp and warm environment of the tropics which results in a quicker and more luxuriant growth of all kinds of plant life. Consequently, in many natural plant communities the space occupied by the vegetation is greater. In the course of evolutionary development many species have reacted to this situation, which is really a struggle for space and light, by producing longer stems. As twining and climbing plants they are able to reach up to lighter and more spacious places.

Climbing plants are an ecologically important group which includes the lianes. Their direction of twining can be clockwise or anticlockwise. The tips of the shoots and parts of the leaves are able to detect suitable supports, usually trunks or branches of other plants though sometimes stems of the same plant may twine together. Some species produce tendrils from their shoots or leaves so that they can attach themselves to their supports, others develop adventitious roots for the same purpose. Climbing plants may be herbaceous or woody, but only the woody ones are called lianes. These are interesting for the curiously shaped structures into which they develop as they grow older.

Lastly, there are species that hook on to their supports by means of thorns and prickles formed in various ways. In all the above groups there are many favourite ornamental plants.

## Bougainvillea

96, 97

*Bougainvillea spectabilis* Willd.

**Four o'clock family** Nyctaginaceae

The Bougainvillea is named after the French navigator Bougainville and is one of the best known and most popular of tropical ornamental plants. When young it can form hedges and at maturity can completely cover the walls of houses and other buildings. The plant is native in eastern Brazil, but is cultivated in gardens as far north as the mediterranean region. It climbs with the aid of the strong, curved spines on its stems, but later on the stems themselves may twine around each other, like Wisteria, and grow to several metres in length.

The leaves are in pairs, and are ovate to lanceolate, entire, and pointed at the end. The really showy part consists of three bracts, arranged in a whorl at the end of the shoots, and variously coloured. In the typical form they are purplish pink, but there are cultivated forms in all shades. The narrowly tubular flower is attached to the base of the bracts and, being relatively small, is overshadowed by them. The pale yellow corolla-tube protrudes from the green calyx and broadens out into five lobes. The seven or eight stamens are enclosed within the tube. The ovary develops into a spindle to pear-shaped fruit. The plant is always conspicuous in flower because of the delicate, net-veined bracts in various colours. They range from a deep rose-purple, through all shades of red and orange to pure white. The Bougainvillea is a plant that loves warmth, and it thrives best in places exposed to full sun.

## Mexican Creeper, Pink Vine, Queen's Jewels

98

*Antigonon leptopus* Hook. & Arn.

**Buckwheat family** Polygonaceae

The Mexican Creeper is a vigorous plant with a woody base that produces loose clusters of beautiful pink flowers. Its stems grow to a length of several metres, climbing with the help of its lateral shoots. The heart-shaped leaves, 10cm long, have wavy edges and long stalks. The flowers have five usually pink perianth-segments, the outer three becoming enlarged at fruiting time. There are seven to nine stamens united into a ring at their base. The ovary, formed from three united carpels, develops into a capsule which remains concealed inside the outer perianth-segments. The Mexican Creeper originates in western Mexico, but is now cultivated everywhere in the tropics, where it frequently escapes and becomes naturalised. It is so well-known that it has acquired numerous names, especially in English and Spanish. Its flowers are often visited by bees. In addition to the typical form, pale pink and white-flowered forms are in cultivation.

## Golden Trumpet

99

*Allamanda cathartica* L.

**Dogbane family** Apocynaceae

The Golden Trumpet is a quick-growing climber from Brazil. It reaches a height of up to 6m and has conspicuous, bright yellow, funnel-shaped flowers 8cm across. The leathery, ovate to oblong, almost sessile leaves grow in whorls of three or four at the base of the slender stems but in pairs higher up. They taper at both ends and have a smooth upper surface. The flowers are produced in small groups at the ends of the stems. The calyx is composed of five sepals, united below into a tube, but with free lobes like small teeth which spread out at the top. The corolla consists of a long narrow tube which broadens out at the end into five rounded lobes. The corolla is typical of the Apocynaceae in being twisted in the bud stage. The mouth of the corolla is fringed with hairs and the five stamens are attached just below this fringe. The ovary, formed from two carpels, develops into spiny capsules which split open into two parts. Inside are numerous seeds with a broad membranous margin. The genus was named after the Dutch botanist Allamand, and the various forms of the species *A. cathartica* include some with the flowers tinged or striped brown.

## Flame Vine 100

*Pyrostegia venusta* (Ker-Gawl.) Miers
Syn. *P. ignea* (Vell.) Presl

**Trumpet-flower family** Bignoniaceae

The Flame Vine climbs with the aid of three-branched, thread-like tendrils, and its generic name means 'fire roof'. Its leathery leaves are composed of two or three leaflets up to 10cm long, shiny on the upper side, elliptic-lanceolate in shape and tapered towards the apex. The underside of the leaf is rust-coloured and the edges are slightly wavy. The brilliant fiery red flowers are arranged in dense panicles at the ends of the stems. The flower has a small calyx and a long, tubular corolla formed from four or five united petals. In bud, the corolla-lobes form a rounded, balloon-like structure over the mouth of the tube, but later roll back slightly as the flower opens. The four stamens and the style protrude from the end of the corolla. The ovary, formed from two united carpels, is inserted into a cylindrical receptacle and later develops into a long, narrow, leathery capsule. Inside are numerous seeds with a narrow membranous margin, arranged in neat rows round a central column. The Flame Vine is a widespread plant, useful as well as ornamental, which can clothe hedges and climb walls. It produces a particularly beautiful effect at flowering-time.

## Chalice Vine, Cup-of-Gold 101

*Solandra nitida* Zucc.

**Nightshade family** Solanaceae

The genus *Solandra* was named after the Swedish botanist Solander who was a pupil of Linnaeus. *S. nitida* is a climbing shrub which is frequently cultivated. It has leathery elliptic leaves with a glossy upper surface. The most striking part of the plant is its flowers. The corolla, composed of five united petals, is tubular at the base, but enlarges abruptly and becomes bell- or cup-shaped, hence the English names for the plant. The flowers are solitary and are cream or pale yellow in bud. During their flowering period, which lasts four days, they change to golden yellow and then to pale orange before they fall. The genus comprises ten species with a natural distribution in Central and South America. Amongst these is *S. grandiflora*, known as Papaturra in Central America. This species produces globular, berry-like fruits, largely enclosed by the calyx, which weigh up to 1kg and have an apple or melon-like taste. It is, however, *S. nitida* which is most widely grown as an ornamental plant in tropical and subtropical regions of the world.

## Sky Vine, Blue Trumpet Vine 102

*Thunbergia grandiflora* Roxb.

**Acanthus family** Acanthaceae

The genus *Thunbergia* is named after the Swedish botanist Thunberg and comprises about 100 species naturally distributed in tropical and southern Africa and Madagascar. It includes a number of decorative plants and can be recognised by the two large bracts at the base of the flower. The most attractive species, cultivated everywhere in tropical countries, is the Sky Vine, which has velvety pale or dark blue flowers. It climbs up walls and fences, and has large, ovate, evergreen leaves, toothed or slightly lobed at the edges, and up to 20cm long. The delicate, handsome flowers hang in small racemes. At the mouth of the corolla, below the five broad, rounded lobes, is a yellowish brown area marked with darker longitudinal lines. The calyx is reduced to a mere rim, and the four stamens are attached just above the base of the corolla-tube. The ovary, formed from two united carpels, develops into a capsule. The plant flowers throughout the whole year.

## Devil's Ivy, Golden Pothos 103

*Epipremnum aureum* (Lind. & André) Bunting
Syn. *Rhaphidophora aurea* (Lind. & André) Birdsey

**Arum family** Araceae

Among climbing ornamental plants of the tropics the genera *Monstera* and *Epipremnum* provide a number of very decorative species. The genus *Monstera* comprises some 25 species native in tropical America, while the natural distribution of *Epipremnum* extends from the Indomalayan archipelago to the islands of the Pacific Ocean. In outward appearance the representatives of both genera look very similar, but *Epipremnum* is distinguished by a one-celled ovary while in *Monstera* the ovary has two or more cells.

The mature form of *Epipremnum aureum* (Greek epi = 'on', premnum = 'tree stump') shown in Photo 103 has variegated yellow and green leaves and has climbed more than ten metres up a kapok tree, *Ceiba pentandra*. *E. aureum* comes from the Solomon Islands. The long, herbaceous, hollow stems attach themselves by means of adventitious roots, many of which form strings round the supporting tree while others hang down like ropes. The same adaptation to environment is shown by the epiphytic species of the genus *Monstera* which includes both erect and climbing species, some of these formerly being placed in the genus *Philodendron*. All the climbing species of both genera are alike in having leaves which, in their juvenile state, have more or less entire blades while older leaves are perforated or split into segments. The formation of holes and division of the leaf are due to an increase in the amount of light.

The inflorescence at the end of the climbing stem is in the form of a club-shaped spadix surrounded by a spathe which is usually greenish on the inside but may also sometimes be yellow or variegated. The spadix of *Monstera deliciosa* (Photo 103, inset) is made up of numerous flowers, each with four to six stamens united with the other flower-parts to form an angular structure. The ovary develops into a berry, and when these berries stand closely packed together they give the appearance of being a single fruit. The fruiting spadix represents a multiple fruit which sheds the remains of the individual flowers when it is fully ripe. The fruits of *Monstera deliciosa* ( = *Philodendron pertusum*) which comes from Mexico and Central America, are aromatic and pleasantly scented, and are used in the preparation of refreshing drinks.

## Cypress Vine, Cardinal Climber, Star Glory 104

*Ipomoea quamoclit* L.
Syn. *Quamoclit pennata* (Desv.) Boyer

**Convolvulus family** Convolvulaceae

The Cypress Vine is an herbaceous annual species with a natural distribution in tropical America. It has twining stems that bear alternate, pinnate leaves, cut to the midrib in linear segments. The narrowly funnel-shaped, crimson flowers are in small, umbellate racemes. They have a short calyx and a slender corolla-tube from which protrude the five stamens. The ovary, formed from two united carpels, develops into a four-celled capsule because of the growth of two false partitions. The colour of the delicate flowers can vary from scarlet through brilliant crimson to purplish red. White and pink-flowered forms are also known. The genus *Ipomoea* comprises 400 species, most of them native in tropical America.

## Pink Trumpet Vine 105

*Podranea ricasoliana* (Tanf.) Sprague
Syn. *Pandorea ricasoliana* (Tanf.) Baill.

**Trumpet-flower family** Bignoniaceae

The genus *Podranea* contains two species, *P. brycei* which is native in Rhodesia and *P. ricasoliana* which comes from South Africa. Because of the beauty of its flowers *P. ricasoliana* is grown throughout the tropics as a climber and trellis-plant. It is a quick growing liane with pairs of pinnate leaves composed of 5–11 leaflets with a toothed margin. Side-shoots are often formed in the axils of the leaves. The pale pink flowers reach a length of 6cm and have dark red stripes in the throat. The narrow corolla-tube, surrounded at the base by an inflated calyx, opens out at the end into a trumpet-shaped structure with five lobes. The ovary, formed from two united carpels, develops into a cylindrical capsule, 25–35cm long. The generic name is an anagram of *Pandorea*, the genus in which it is sometimes included, and which was given because of the similarity of the capsule to Pandora's box. This splendid decorative vine has no special soil requirements but thrives best in full sun. It climbs rapidly up walls and over pergolas, and is a favourite plant for covering summer-houses.

## Balsam Pear 106

*Momordica charantia* L.

**Gourd family** Cucurbitaceae

The Balsam Pear was originally found only in the tropics of the Old World, but it has been spread by man throughout all the tropical regions of the world. It is an annual herbaceous plant which climbs up to a height of 2m with the aid of undivided tendrils formed from modified shoots. Opposite the tendrils are the circular or kidney-shaped leaves which have five to seven lobes, ovate to oblong in shape and sometimes toothed along the margin. The plant produces small, regular, five-lobed male and female flowers of an orange-yellow colour. The inferior ovary of the female flower develops into an oblong fruit with a somewhat bumpy surface. The fruit, a berry, hangs down on a slender stalk and is green at first but later turns bright orange. Inside are the pale grey to brown, slightly flattened seeds with a raised pattern on both sides. They are surrounded by a blood red pulp which forms a striking contrast with the orange-coloured skin of the fruit. Also common in tropical regions is the closely related Balsam Apple, *M. balsamina*. This has broadly ovoid fruits with a ridged surface. Its leaves are circular in outline, cut halfway towards the centre into three to five lobes, and are used in the preparation of a tea and as a tonic.

## Glory Bower **107**

*Clerodendrum splendens* G.Don

**Verbena family** Verbenaceae

The name *Clerodendrum* is derived from the Greek kleros (chance) and dendron (tree), since in this genus some species are beneficial while others have a harmful effect. The genus comprises 390 species mainly distributed in the Old World. *C. splendens* is a climbing shrub from the mountains of West and Central Africa, and can reach a height of 10m on a suitable support. The woody shoots bear pairs of shortly stalked, narrowly heart-shaped, pointed leaves with a wavy edge. The flowers are grouped in umbellate panicles at the end of the shoots or in the leaf-axils. They have a small calyx and a corolla with a long narrow tube, purplish or more often bright scarlet in colour, from which the four stamens and the long style project. The two-celled ovary develops into a berry-like drupe. This plant is one of the most beautiful of tropical climbers. *C. thomsoniae*, a related species found in West Africa and the Congo, is also worth mentioning. It is now equally widespread in tropical regions as a decorative plant. It has a large, yellowish white, inflated calyx, which forms a striking contrast with the velvet-red corolla. Both species are frequently grown in glasshouses in temperate latitudes.

## Ipomoea **108**

*Ipomoea carnea* Jacq.

**Convolvulus family** Convolvulaceae

The genus *Ipomoea* comprises 400 tropical and subtropical species, most of which are native in America. They include erect shrubby or tree-like forms, but are predominately annual or perennial climbers which cover the bushes and hedges of the American tropics with masses of flowers. Those with blue or purple flowers are especially beautiful. The flowers have a velvety sheen and are produced either in the axils of the alternate leaves, or in clusters at the ends of the shoots. They have a more or less short corolla-tube and five stamens, usually of different lengths. The thread-like style ends in a globular or sometimes divided stigma. The ovary, formed from two united carpels, develops into a capsule which may split into two or four parts, or may open by means of a lid. Usually four, but occasionally six large seeds are produced.

### Blue Passion-flower

109

*Passiflora caerulea* L.

**Passion-flower family** Passifloraceae

The genus *Passiflora*, Passion-flower, comprises 400 species mainly distributed in tropical America, though there are also representatives in Asia and Australia, as well as in New Caledonia, Fiji, New Guinea, and one species in Madagascar. Most species climb by means of tendrils on their stems. Their flowers, which arise singly or in pairs near the tendrils, may be inconspicuous or very striking. They always have the same characteristic structure which has led to the name 'Passion-flower' being given to the genus. At the base of the flower are three leaf-like structures, one bract and two bracteoles, which in a way represent the calyx since the five sepals have a similar colouring on their inner surface to the five petals. This gives the impression of a ten-petalled flower. Next comes the corona. This is thought to be formed from the enlarged top of the flower-stalk and consists of an inner and outer ring of fine filaments. Sometimes this occurs as a membranous folded or curled ring. In the centre, rising up above the corona, is a column composed of the five stamens and the three-carpelled ovary with spreading styles. Nectar is secreted at the base of this column. The plant received its name because it reminded its discoverers of Christ's crucifixion: the corona was the crown of thorns, the five stamens were the wounds, and the three styles, broadened at the ends into club-shaped stigmas, represented the nails.

The Blue Passion-flower comes from Brazil and Peru, and is also grown as a house-plant. It is, however, a vigorous climber with five to seven-lobed leaves, rather kidney-shaped stipules, and pleasantly scented flowers up to 8cm in diameter, having white or pinkish sepals and petals and a blue corona. Another species often cultivated as a decorative plant is *P. racemosa* which has loose, pendulous racemes of scarlet flowers.

# Herbaceous Plants

## Coastal Plants

### Beach Morning Glory

110, 112

*Ipomoea pes-caprae* (L.) Roth

**Convolvulus family** Convolvulaceae

The Beach Morning Glory is found on all tropical coasts, especially on sandy beaches. Like the Sea Bindweed, which occurs on mediterranean and atlantic shores, it forms long, creeping stems. Under favourable conditions these can reach a length of 10m or more. Their numerous adventitious roots serve to stabilise sandy coastal areas. The creeping stems grow to a considerable length along the ground and bear stalked, almost circular leaves which are usually notched at the apex to a greater or lesser extent, reminiscent of the imprint of a goat's foot – hence the name of the species 'pes-caprae'. The large, pink, trumpet-shaped flowers (Photo 112) may be more than 10cm across, and, like the whole plant, are very striking. The two-celled ovary develops into a large capsule with two seeds in each cell. The seeds can germinate and grow in sea water, and this has contributed to the world-wide spread of the plant. The small pink flowers and fresh green leaves in Photo 110 belong to the Seaside Bean, *Canavalia rosea* (*C. maritima*), a member of the Papilionaceae often found growing in association with *Ipomoea pes-caprae*.

## Giant Milkweed

**111**

*Calotropis procera* (Ait.) Ait. f.

**Milkweed family** Asclepiadaceae

The Giant Milkweed is naturally distributed from West Africa to India, and often occurs abundantly in steppe and desert regions. In India it is known as Madar or Yercum and its distribution extends westwards in Asia as far as the Dead Sea. The plant becomes woody, especially at the base, and may grow to more than 4m in height. In the Sahara it may even become a small tree. The stout, erect stems bear large, broadly ovate, pointed leaves covered with a dense layer of hairs. The leaves are shortly stalked and are arranged in pairs along the stems. The veins are coarse and conspicuous, particularly on the underside of the leaf. The flowers are formed in the axils of the leaves and the corolla-tube opens out into five lobes. The corolla is mainly white, with purple areas on the inside of the lobes. The ovary, formed from two carpels, develops into a large, egg-shaped, green fruit tinged with red which is known as Apple of Sodom. It contains numerous seeds with plumes of fine hairs which provide a 'vegetable silk'. They are not very suitable for weaving so are mainly used as stuffing material. The genus comprises five species, native in the warm regions of Africa and Asia. A few decades ago *C. procera* was introduced into tropical America and in a short time spread over large areas. In southern Asia and Malaysia, the bark of the related species *C. gigantea* provides the high quality Yercum fibre. The milky juice of this species resembles gutta-percha in its dry state, and the hairs on the seeds can be used for spinning.

## Butterfly Pea

**113**

*Centrosema* sp.

**Pea family** Fabaceae

The genus *Centrosema* comprises some 30 species which occur wild in the tropical and temperate regions of America. It gets its name from the short spur which can be found on the back of the standard or upper petal of the flower. It is a climbing or sometimes creeping plant with pinnate leaves, usually consisting of three relatively large leaflets. The attractive flowers are on fairly long stalks which arise from the axils of the leaves. The flowers have a very broad or even heart-shaped standard petal and the two wing petals are bent inwards so that they lie alongside the keel. The single carpel develops into an almost sessile, linear pod. In the Caribbean area and tropical America, *C. pubescens*, a species with carpels hairy on the underside, is widely distributed. The closely related *C. virginianum* is found also in the tropical and subtropical regions of America and also in Virginia. This has a pod 5–7mm broad, whereas that of *C. pubescens* has a breadth of only 3–5mm.

# Perennials

## Red Ginger 114, 115

*Alpinia purpurata* (Vieill.) Schumann

**Ginger family** Zingiberaceae

The genus *Alpinia* comprises 230 species, and is named after the Italian botanist Prospero Alpino who lived round about 1600. It is a handsome plant with an aromatic tuberous rootstock that is often branched. The stems which grow up from the rhizome bear alternate, lanceolate leaves somewhat like those of the Banana. The Red Ginger originates in Malaya but is now widespread in tropical gardens as an ornamental plant. It reaches a height of 1.5–2m and has brilliant red spikes of flowers throughout most of the year (Photo 115). The decorative effect of the plant is due to the fleshy bracts which are wax-like and glossy. In the axils of these bracts pure white flowers arise which are less striking. Monocotyledonous plants normally have six stamens, but, like all members of the Ginger family, the Red Ginger has only one fertile stamen, the centre one of the inner whorl. The other two stamens of the inner whorl are united to form a lip or labellum. The ovary, formed from three united carpels, develops into a capsule which remains closed long after it is ripe. It is characteristic of the plant that, like the Agave, it produces young plants while it is in flower. These grow on the mother-plant until they are about 30cm high and in this way the plant easily reproduces itself.

## Shell Ginger 116

*Alpinia zerumbet* (Pers.) Burtt & Smith
Syn. *A. speciosa* (Wendl.) K. Schum.

**Ginger family** Zingiberaceae

The Shell Ginger is native in China and southern Japan and reached Europe via India as early as 1792. It is a decorative perennial plant that is cultivated in all tropical gardens. It grows to a height of up to 3m and has stout stems which bear lanceolate leaves shortly pointed at the apex and gradually narrowed towards the base. Like the leaves of the Red Ginger they are smooth and without hairs. The very decorative inflorescence is pendulous and reaches a length of 30cm. It consists of numerous flowers densely crowded together. The large sepals are white, like the petals, and are united to form a bell-shaped calyx. The lip, formed by the fusion of the two inner infertile stamens, is up to 4cm long. It is broadly ovate in shape, three-lobed and the edge is rolled over. It is yellow with red spots and stripes. The fruit is a red capsule. As in all Zingiberaceae, the seeds have a conspicuous appendage.

The aromatic rhizome of several species of *Alpinia* is used for medicinal purposes and for flavouring. The best known is the Galangal, *A. officinarum*, which comes from the peninsula of Hainan in southern China and is also cultivated in Thailand. Its rootstock reaches a length of one metre and yields a drug with a spicy smell and a bitter, slightly burning taste. It contains 0.5–1.5% of ethereal oils, also galangol and alpinol, two sharp tasting substances.

The Greater Galangal, *A. galanga*, which is native in the Moluccas, produces a rhizome that is used for medicinal purposes just as the Galangal is used in southern Asia.

## Garland Flower, Butterfly Ginger **117**

*Hedychium coronarium* Koenig

**Ginger family** Zingiberaceae

The genus *Hedychium* comprises 50 species most of which are native in India, but two originate in the Philippines and two come from Madagascar. They are usually erect plants with a tuberous rootstock. The leaves have a much reduced stalk but the flower-spike is long-stalked. Probably the best known of all the species is *H. coronarium*, which grows wild in the Himalaya and reaches a height of more than 1m in cultivation. It has stiff, oblong-lanceolate, finely pointed leaves, smooth on the upper side but softly hairy beneath. They grow up to 60cm long and 30cm broad. The large, snow-white flowers are arranged in few-flowered spikes up to 20cm in length, and their bracts are green with a brownish margin.

The large, strongly scented flowers have at their base three sepals united to form a long, narrow tube. Then come three petals, united below to form the corolla-tube, and separating above into three narrow, pointed lobes. Two of the three stamens are infertile, petal-like, and are joined together to form the strikingly large lip, which is roughly heart-shaped with a slit at the top. The single fertile stamen wraps closely round the style, forming a long, thin structure that protrudes in a strange way far beyond the rest of the flower. The stigma is covered with sticky hairs and forms the tip of this curious structure. In the evenings the white flower gives out a pungent scent, which is why the genus was named *Hedychium*, the Greek for 'sweet snow'.

## Crape Ginger **118**

*Costus speciosus* (Koenig) Smith

**Ginger family** Zingiberaceae

In Greek, the word 'Costus' means a spicy, pepper-like root, and in Latin it refers to an Indian shrub, from the root of which a valuable ointment was made. Like many members of the Ginger family it contained aromatic substances, which are also present in the Crape Ginger, *C. speciosus*, native in India but now cultivated everywhere in tropical gardens. The strongly aromatic rootstock gives rise to a number of erect stems, semi-woody at the base, and up to 3m in height. The numerous lanceolate, shortly stalked leaves are silky hairy, up to 20cm long, and are arranged spirally on the stems. The leaves taper gradually at the apex and there is a characteristic ligule at the base. The flowers are densely arranged in erect, elliptic or egg-shaped spikes and have red bracts. The sepals, like the petals, are united below into a tube. The really showy part of the flower is produced by the five stamens which are joined together to form the 'lip', so called because of its shape. This is white in colour with an orange-red centre. The three-carpelled ovary develops into a capsule. The species is very variable and sometimes has stems which are branched in the upper part. Another species often cultivated is the Spiral Ginger, *C. afer*, which is native in Senegal and Lagos. Its leaves are slightly smaller and are drawn out into a tail-like tip. The white lip has a rather irregularly toothed edge and is yellow at the base.

## Kahili Ginger 119

*Hedychium gardnerianum* Rosc.

**Ginger family** Zingiberaceae

Like the Butterfly Ginger, Kahili Ginger is also native in the Himalaya and northern India. Its flowering stems reach a height of more than 1.5m and end in a loose spike of flowers. The leaves are linear to lanceolate and are drawn out into a fine point. They are more than 45cm long and 15cm broad. The golden yellow flowers grow in pairs in the axils of the bracts. They are similar in structure to those of the Butterfly Ginger but are somewhat narrower. They have a two-lobed lip and red stamens which are twice as long as the corolla. The strong scent of the flowers is reminiscent of oranges or jasmine. The ovary, formed from three united carpels, develops into a red capsule containing seeds surrounded by an aril or extra covering.

## Wild Plantain 120

*Heliconia wagneriana* Petersen

**Banana family** Musaceae

The genus *Heliconia* is named after Mt Helicon, the abode of the Greek muses in Euboea, and comprises 150 species native in tropical Central and South America. All species have rhizomes from which arise false stems, formed from the rolled leaf-sheaths, as occurs in the Banana. Several species reach a height of more than 3m, and about all have large, long-stalked, lanceolate leaves arranged in two rows. Like many of the Musaceae the leaves tear parallel to the lateral veins that run out at right-angles to the midrib giving the impression of a pinnate leaf. The inflorescence is striking on account of the large, boat-shaped, orange to bright red bracts, which conceal in their base the relatively inconspicuous group of flowers. The typical flower has three sepals, three petals, and five fertile stamens. The sixth stamen is infertile and petal-like. The ovary, formed from three united carpels, develops into a capsule which splits into three when ripe, one seed in each part. All species of *Heliconia* produce an abundance of sweet, sticky nectar at the base of the petals. This oozes out into the bracts and becomes mixed with any rain water that has collected there. The flowers are well adapted to pollination by birds.

The very showy species *H. wagneriana*, Wild Plantain, comes from Panama and reaches a height of up to 2.5m. It usually has four or five long-stalked, linear-lanceolate leaves, tapered at the apex, which may grow to be more than 80cm in length. The broad, boat-shaped bracts are arranged in two rows, on either side of the stem, and form a compact inflorescence. They are deep pink in colour, shading to orange or yellow towards the margin, which is marked with a green line. The upper bracts are shorter and more crowded than the lower, and are less finely pointed. A considerable amount of water collects in the bracts. One of the three sepals is free and distinctly enlarged, but the other two sepals and the three petals are joined together. The flower is typical in having five fertile stamens and one short, infertile, petal-like stamen. Each flower produces a large amount of nectar.

## Parrot's Plantain 121

*Heliconia psittacorum* L.

**Banana family** Musaceae

The Parrot's Plantain is native in Guayana and Brazil and grows up to 1m high. It has long-stalked leaves, up to 45cm long and 5cm broad, narrowed at the base into the stalk. The inflorescence is coloured in shades of red and orange producing an exotic, iridescent effect, which gave the plant its name. It is often possible to see the transition from leaf to bract in the bracts of the inflorescence. The inflorescence grows up to 9cm in length and has four or five brilliant red bracts. The flowers themselves are greenish yellow in colour and have black spots near the tips. The plant can vary greatly in size and colour.

## Beaked Heliconia 122

*Heliconia rostrata* Ruiz & Pavon

**Banana family** Musaceae

This species comes from Peru and is one of the most beautiful in the genus. It is a stout plant, over 3m high, and has leaves 2–3m long arranged in two rows up the stem. The leaves have a round stalk and a linear-lanceolate blade, ending in a short point. Both blade and stalk are slightly glaucous. The stout, raspberry-red stalks bearing the decorative inflorescences bend over so that the flowers are pendulous. Each inflorescence has up to 12 short, bluntly pointed bracts which stand apart from each other when the plant is in full flower. The bracts are a brilliant red at the base shading to yellow at the tip with a green edge, and resemble in shape a bird's curved upper beak. The flowers are a bright sulphur yellow and poke out of the bracts like the beaks of fledglings from a nest.

## Bird-of-Paradise Flower 123

*Strelitzia reginae* Banks

**Banana family** Musaceae

The genus *Strelitzia* was named after Charlotte von Mecklenburg-Strelitz, Queen of George III, and comprises four species in South Africa. By far the commonest and best known species is the Bird-of-Paradise Flower, *S. reginae*. From the stout rhizome arise leaves and inflorescences which grow to about the same height of 1–2m. The leaves are similar to those of the Banana and consist of a long, stout stalk and a smooth, leathery, oblong-ovate blade with a usually reddish midrib. The highly decorative inflorescence appears amongst the leaves. There is a boat-shaped bract, often tinged purple, from which the flowers appear, opening in succession from the base in the course of the flowering period. The individual flowers consist of three outer perianth-segments, orange in colour and narrowly lanceolate in shape, and three inner segments of a beautiful metallic blue. Two of these are united to form an arrow-shaped sheath enclosing the five fertile stamens and the style. The ovary, formed from three united carpels, develops into a many-seeded capsule.

## White Bird-of-Paradise Flower 124

*Strelitzia augusta* Thunb.

**Banana family** Musaceae

In contrast to *S. reginae*, the White Bird-of-Paradise Flower, native in the Cape Province and Natal, has a clearly defined stem. The stalk of the banana-like leaves can be as much as 2m in length, while the blade, heart-shaped at the base, is up to 1m long. The colour of the boat-shaped bract is variable. In some cultivated forms it is a metallic blue, in others reddish purple. The three outer perianth-segments are milky-white in colour, narrowly lanceolate in shape and long-pointed. As in *S. reginae*, two of the three inner perianth-segments are modified to form an apparently homogeneous arrow-shaped structure which encloses the five stamens and the style. The three-celled ovary develops into a capsule containing numerous seeds. Besides the White Bird-of-Paradise Flower, the two other species in the genus are in cultivation.

## Snow Flower 125

*Spathiphyllum floribundum* (Lind. & André) N.E. Br.

**Arum family** Araceae

Of the 36 known species in the genus *Spathiphyllum*, by far the majority originate in the American tropics. Only two species come from the Indomalayan archipelago, including *S. commutatum* from the Philippines. Most species have a white spathe, and identification is often made more difficult because of hybridisation. The best known is *S. floribundum* which is native in Colombia. The plant has a short stem, and both leaf-stalk and blade are about the same length. The stalk is 12–20cm long, clearly sheathing at the base, and the blade is oblong-ovate and pointed, lustrous green on the upper side and dull beneath. The main veins are scarcely visible on the upper surface of the leaf-blade but are prominent on the underside. The spathe is snow-white and tapers at the base into the stalk which is 20-30cm long. It envelops the base of the spadix, then broadens out, forming a characteristic angle with the spadix. It remains white for a long time and only turns a greenish colour as it withers. The spadix has no sterile area, but is densely covered with bisexual flowers which later produce berries containing a few seeds. The genus was given its name because of the leaf-like spathe (phyllon = leaf). It is characteristic of the genus that the spadix is shorter than the spathe.

## Arum Lily, Calla Lily, Trumpet Lily 126

*Zantedeschia aethiopica* (L.) Spr.

**Arum family** Araceae

The genus *Zantedeschia* is named after the Italian botanist and physicist F. Zantedeschi and comprises eight species, all of them originating in South Africa. The best known is *Z. aethiopica*, which, contrary to the name of the species, comes from the flat, damp or swampy areas in the neighbourhood of Cape Town, and not from Ethiopia. Outside the tropics, it is often grown as a house-plant in soil or shallow water. It has a thick, fleshy, poisonous rhizome from which arise the large, handsome leaves. The long leaf-stalks, sheathing at the base, open out into leathery, arrow or heart-shaped blades. The blade has a coarse midrib with lateral veins that extend to the edge of the leaf. The inflorescence rises above the leaves, reaching a height of more than 1m.

Its most conspicuous part is the white spathe, 12–16cm long, which is tubular at the base and funnel-shaped above, spreading out into a flat, pointed brim at the top. The spathe envelops the spadix, which has female flowers at its base and male flowers above. The egg-shaped ovaries develop into berries, like those of our native Lords-and-Ladies or Cuckoo-pint. It is interesting that the ovaries are not receptive to pollen from the same plant or from its offsets. The Arum Lily often becomes naturalised in the tropics, also in Madeira and the Canary Islands. In temperate regions a smaller form, with an inflorescence only 0.5cm high, is grown as a house-plant. There are a number of cultivated varieties in existence. *Z. aethiopica* is rather tender, but the smaller, spotted-leaved species *Z. albo-maculata* can be grown outside in sheltered places. Other species which are used as ornamental plants are *Z. melanoleuca*, *Z. pentlandii*, and *Z. rehmannii*.

## Crystal Anthurium **127**

*Anthurium crystallinum* Lind. & André

**Arum family** Araceae

The Crystal Anthurium is native in Colombia and has large, very decorative leaves and a short stem. The leaf-stalks are long, round or sometimes slightly winged, and the dark green blades, broadly heart-shaped and with overlapping basal lobes, have a velvety sheen. The principal veins are marked out in silver on the upper side of the leaf-blade. The blade is up to 50cm long and 30cm broad, pale green on the underside and tinged with pale pink in its young state. The erect, dark brown flower-stalk extends above the leaves, bearing a narrow, brownish, membranous spathe, often turned back at the edges, which is so inconspicuous that it appears to be only rudimentary. From its base grows the slender, greenish, cylindrical spadix. Like many species of *Anthurium* that have their home in the damp rain-forests and mountain regions, this plant needs conditions of shade and humidity. The undoubted decorative value of the plant lies in its ornamental leaves.

## Anthurium hybrids **128**

*Anthurium x cultorum* Birdsey
Syn. *A. andreanum* hort. non Linden

**Arum family** Araceae

The parents of this very varied group of hybrids cannot now be ascertained. The hybrids differ considerably in size, as well as in the form of the leaf and the spathe, from the little Flamingo Flower, *A. scherzerianum*, which occurs wild as a pure species in the Amazon region, and has a slender, curved or spirally twisted spadix. *A. x cultorum* has long-stalked, narrowly heart-shaped leaves up to 30cm long. The blade ends in a distinct point and is dull green in colour. The spathe spreads out wide and is very glossy, almost as if it had been varnished. There is a distinct network of veins on the upper side and often a short, blunt point. The veins are often swollen and the colour of the spathe varies from dark to light red and salmon-coloured to white. The spadix always grows upright and is straight or only slightly curved. The flowers set fruit without being fertilised, and, as the spadix becomes more fleshy with age, the fruits protrude from its surface. As is the case with the spathe, the spadix may be flecked with green due to the formation of chlorophyll.

## Medinilla 129

*Medinilla magnifica* Lindl.

**Melastoma family** Melastomataceae

The genus *Medinilla* is named after the Spanish governor of the Marianas Islands, José de Medinilla y Pineda, and comprises 300 species of erect or epiphytic shrubs, or, more often, root-climbers. Their distribution extends from West Africa via Indonesia to the Fiji Islands. Probably the most impressive species is one which is native in the Philippines, *M. magnifica.* This is a shrub which grows up to 1.5m in height and has thick, four-winged stems with tufts of bristles at the nodes. The leathery leaves are arranged in pairs, and, as they have no stalks, the base folds slightly round the stem. The leaves are up to 30cm long, usually heart-shaped or ovate in shape but sometimes oblong, and they end in a fine point. The three strong veins, branching upwards from the base, are characteristic of the family. The numerous pink flowers are arranged in pendulous, terminal or axillary panicles, up to 50cm in length, and are enhanced by very large, pinkish white bracts. The ovary develops into a fleshy berry. Outside the tropics, this species is one of the most beautiful plants for the warm greenhouse.

## Calathea 130

*Calathea cylindrica* (Roscoe) K. Schum.

**Arrowroot family** Marantaceae

The genus *Calathea* comprises 130 species from the humid regions of tropical America, and its name is derived from the Greek word 'kalathos', a basket, because the lip of the flower is basket-shaped. This strong-growing plant is native in Brazil and reaches a height of more than 1.5m. The leaves have a long stalk, and a smooth, thin blade, up to 50cm long, and elliptic or oblong in shape with a short point. The cylindrical inflorescence is hidden amongst the foliage, and the most striking parts are the fleshy, pale green bracts, arranged in a spiral round the stem. The flowers appear in the axils of the bracts, and the perianth consists of a slender, tubular corolla which opens out into three lobes, and a calyx of three small, greenish white sepals. The three corolla-lobes are tinged green on the outside. The genus *Calathea* contains a number of species with attractively marked leaves, including *C. makoyana*, also from Brazil, which has leaves patterned with cream and green above and purple beneath, and is perhaps the most striking plant of all.

## Desert Rose 131

*Adenium obesum* (Forsk.) Roem. & Schult.

**Dogbane family** Apocynaceae

The genus *Adenium* comprises ten species distributed in tropical and South Africa, also in Arabia. Most of them are steppe and desert plants with thick, fleshy stems. The best known is the Desert Rose, *A. obesum*, which comes from the dry parts of East Africa. This species has succulent stems which later become woody. The narrowly obovate, shortly pointed leaves are arranged spirally on the stems. The attractive pink flowers are twisted in bud, then open to reveal a long corolla-tube and five widely spreading lobes. An abundance of milky juice oozes out of the plant if any part of it is damaged. This is highly toxic and has been used as an arrow-poison.

## Shrimp Plant 132

*Beloperone guttata* Brandeg.

**Acanthus family** Acanthaceae

The genus *Beloperone* comprises 30 species originating in the warmer regions of the New World. The Shrimp Plant comes from Mexico and is cultivated everywhere in the tropics. It is a semi-shrubby species growing up to 1m in height, with numerous branches covered in short hairs. The leaves are 2–6cm long, shortly pointed at the apex, and narrowing abruptly at the base into a thin stalk. They are without teeth on the margin but have short hairs on both sides. The arching spikes of flowers are 10–20cm in length and are particularly interesting on account of the softly hairy, brownish red bracts which overlap each other like roof-tiles. From the axils of these bracts arise the white, slightly curved flowers which soon wither and fall. The corolla is clearly divided at the end into an upper and a lower lip, the latter marked with three rows of reddish purple dots towards the base. The ovaries develop into long-stalked capsules.

## Pachystachys 133

*Pachystachys lutea* Nees

**Acanthus family** Acanthaceae

The genus *Pachystachys* is very closely related to *Beloperone* and comprises seven species, also from the American tropics. It is a very decorative plant with angular stems which reach a height of 40–50cm, and bear pairs of lanceolate leaves, shiny on the upper side and prominently veined. The most conspicuous parts of the inflorescence are the bright yellow, heart-shaped bracts, arranged precisely in four rows. The white flowers, which arise in the axils of these bracts, resemble the family Labiatae as the corolla has a distinct upper and lower lip. Inside the corolla-tube are four stamens and a superior ovary, formed from two united carpels, which develops into a capsule. In temperate regions this plant can be grown in a greenhouse.

## Good-luck Plant, Tree-of-Kings 134, 135

*Cordyline fruticosa* (L.) A.Chev.
Syn. *C. terminalis* (L.) Kunth

**Lily family** Liliaceae

There are about 20 species in the genus *Cordyline* but *C. fruticosa* is especially well-known for its variety of form and is found as a decorative plant everywhere in the tropics. Its natural distribution extends from India, through the Indo-malayan archipelago and northern Australia to New Zealand and Papua. It is a small shrub with a slender, simple or branched stem reaching a height of 4–5m. The leaves are 30–50cm long, lanceolate in shape, and may be green or brightly variegated, usually however red. When they fall, characteristic scars are left on the trunk. The inflorescence hangs over on one side and is made up of a loose cluster of whitish, lilac, or reddish coloured flowers about 1.5cm long. The flowers are followed by red berries. In temperate latitudes the plant is often grown as a pot-plant in glasshouses and then takes the form of a half-shrub (Photo 134). The origin of the many forms and varieties can now no longer be ascertained.

## Boat Lily

136

*Rhoeo spathacea* (Sw.) Stearn
Syn. *R. discolor* (L'Hérit.) Hance ex Walp.

**Spiderwort family** Commelinaceae

The genus *Rhoeo* contains only one species which originated in Central America. From there it was spread by man throughout the tropics and quite often becomes naturalised. In gardens it is frequently used as an edging to paths. It is a plant with short, stout stems which end in a tuft of linear-lanceolate leaves, up to 35cm long and 8cm broad, directed upwards at an angle. They are usually reddish purple on the underside. The white flowers, composed of three petals and three sepals, are crowded amongst mussel-shaped bracts produced in the axils of the leaves. Apart from the typical form, with a green upper surface to the leaf, there are variegated forms with yellowish stripes. The best known of these are 'Vittata' and 'Variegata'. The cells on the lower surface of the leaves owe their colouring to the pigment anthocyanin. It should be pointed out that the whole plant is poisonous.

## Crinum

137

*Crinum augustum* Roxb.

**Daffodil family** Amaryllidaceae

The genus *Crinum* comprises 150 species in Asia and Africa and includes a number of attractive coastal plants. One of the most striking is *C. augustum*, native in Mauritius and the Seychelles, which appears to be a slightly smaller form of *C. amabile* from Sumatra. In fact *C. augustum* has been considered by some authorities to be merely a variety of this species. Like all members of the genus, *C. augustum* produces a large bulb with a distinct neck. The leaves are numerous, up to 1m long and 10cm broad, and pointed at the end. The stem is flattened to the extent of being two-edged and at the top, just above a pair of bracts, is an umbel of decorative, sweetly scented flowers tinged with a wine-red colour. The large flowers have an inferior ovary, and a long corolla-tube which opens out into six widely spreading lobes, coloured red on the outside, like the stamens. The ovary, formed from three united carpels, ends in a capitate stigma, and develops into an almost spherical capsule which splits irregularly into several parts.

Another widely cultivated species is *C. asiaticum* which is very variable and has a large number of different forms.

## Candelabra Aloe

138

*Aloë arborescens* Mill.

**Lily family** Liliaceae

The genus *Aloë* comprises 250 species which are found in the dry regions of East Africa from Ethiopia to Cape Province and on Socotra and the Mascarene Islands which lie to the east of the main area. The Candelabra Aloe comes from South Africa and grows up to 3m high with a branched stem about 8cm in diameter. The rosettes of leaves at the ends of the branches are up to 1m across. The sword-shaped leaves are set closely together, some being erect and others spreading. They are 45–60cm long, 5–7cm broad and up to 2cm thick, dull grey to bluish green in colour with triangular spines up to 5mm long. The plant is very variable in form and the inflorescence may reach a length of 80cm. The flowers may be narrowly or broadly conical in shape and can vary considerably in size. Their stamens are concealed within the corolla-tube. The three-celled ovary develops into a capsule. Numerous varieties of this species have been described. Other species are the source of 'bitter aloes', a drug prepared from the juice of their leaves.

## Scarlet Plume **139**

*Euphorbia fulgens* Karw. ex Klotzsch

**Spurge family** Euphorbiaceae

The genus *Euphorbia* is large and varied. It comprises 1600 species, both herbaceous as well as woody, also plants which are succulent in varying degrees. A characteristic of all species is the poisonous milky juice which is produced in the long, branching cells known as latex tubes. One of the species which is most popular in temperate zones as a decorative greenhouse plant or for cut flowers is *E. fulgens*. It is a perennial plant which grows slightly woody with age, and its slender, almost unbranched stems, bearing alternate leaves, are without thorns or hairs. The leaf-blade is lanceolate with an entire margin, long-pointed at the apex, and tapering at the base into the stalk. It varies between 6–14cm in length and 1.5–2.5cm in breadth and has a stout midrib and thin lateral veins.

The ornamental character of this plant is enhanced by the attractive inflorescences (cyathia) which are produced at the end of the stem, all facing in the same direction. Each cyathium has a very complex structure in which the red or orange appendages of the five glands are arranged in a circle and give the impression of true petals. Below each appendage is a transverse nectary, and in the centre of the cyathium is a group of individual flowers, modified to such an extent that the male flowers are represented only by stamens and the single female flower consists of an ovary on a short, curved stalk. The ovary, formed from three united carpels, develops after fertilisation into a small capsule which splits into three parts. The main flowering-time of this highly decorative plant is in December and January.

## Crown-of-Thorns **140**

*Euphorbia milii* Desm.
Syn. *E. splendens* Boj.

**Spurge family** Euphorbiaceae

The Crown-of-Thorns, which comes from Madagascar, is without doubt the best known species in the genus *Euphorbia*. In temperate climates it is grown as a pot-plant for the window-sill or the greenhouse, while in the tropics and subtropics it is often found outside as an individual shrub or in hedges. It is especially well suited as a hedge-plant because it can grow to a height of 2m and is formidably armed with sharp thorns. The slightly angled, branched stems are flexible and intertwine easily. The bright green leaf-blades are bluntly ovate to oblong-spathulate in shape and up to 8cm in length, and the base of the blade narrows into the stalk. The leaves fall quickly and leave behind a scar between two small cushions on the shiny, bright green stems. On each side of the scar, or the leaf-stalk if it is still there, are pairs of thin, sharp thorns, up to 2cm long, which represent modified stipules. They cover the entire length of the stem and give rise to the common name of the plant. The flowers appear, rather like those of the Scarlet Plume, in cyathia between the young thorns at the end of the stems. The inflorescence is repeatedly branched, with a pair of cyathia at the end of each branch, each cyathium being surrounded by two bright red, shortly pointed bracts. These bracts form the showy part of the inflorescence. In the axils of the bracts are the individual flowers, modified in a similar way to those of *E. fulgens*.

The Crown-of-Thorns is a very variable species, and the various cultivated forms have been given distinguishing names. *E. milii*, which flowers mainly during the winter and spring, is distinguished by having its thorns slightly enlarged at the base. The closely related *E. bojeri* is more succulent, has obovate to wedge-shaped leaves, and almost circular, bright red bracts round the cyathium. There is a variant of *E. milii* with pale yellow bracts called 'forma *lutea*'. As the Crown-of-Thorns can stand drought it is a favourite plant for cultivation in the hot, dry regions of the tropics.

## Blood Flower 141

*Asclepias curassavica* L.

**Milkweed family** Asclepiadaceae

The genus *Asclepias* is named after Asclepios, the Greek god of medicine, and comprises 150 species native from the tropics of South and Central America northwards to Mexico and the United States. The original distribution of *A. curassavica* was in the American tropics but because it is an attractive plant it has been spread by man to all the warmer lands and frequently becomes naturalised. It is of semi-shrubby growth and reaches a height of over 1m. Its usually unbranched stems bear shortly stalked, lanceolate leaves, 12–15cm long, dark green above and bluish green beneath. The inflorescence is in the form of a loose umbel of five to ten flowers. The individual flowers are about 1cm across and have only a small calyx. The five corolla-lobes are deep orange-red, and the five corona-hoods, outgrowths from the stamens, are bright orange and form the centre of the flower. These corona-hoods are ovate in shape and flattened at the sides. The ovary consists of two free carpels with their styles also free but with the stigmas joined to the stamens. Each carpel develops into a dry fruit which splits down one side to release the numerous seeds, each with a tuft of silky white hairs. Only a few of the Asclepiadaceae provide hairs suitable for spinning.

## Four-o'clock 142

*Mirabilis jalapa* L.

**Four-o'clock family** Nyctaginaceae

The name of the genus is taken from the Latin word for 'wonderful' and was given because of the marvellous range of colours of the flowers. The colouring of this species can be white, pink, red, yellow, yellow striped with red, or white striped with red. There are even flowers with intermediate shades involving three or four colours. Because the species is so variable it has been used for research into the principles of heredity.

The plant is normally a perennial growing 60–100cm high, and comes from Mexico. It has blackish, tuberous roots and branched stems that are often tinged red at the nodes. The leaves are hairless, usually in pairs but sometimes also alternate, ovate-lanceolate in shape, with an almost heart-shaped base and a long-pointed apex. The flowers are in clusters at the ends of the stems. At the base of the flower is a small, green involucre like a five-toothed calyx. Above this is the true calyx which is corolla-like in appearance and has a long, slender tube opening out into a funnel at the top. The tube is constricted just above the one-celled ovary. The capitate stigma and the five stamens protrude from the top of the coloured, funnel-shaped calyx. The superior ovary develops into a capsule, and as this grows larger the surrounding calyx enlarges also. The tuberous roots were formerly used as a strong purgative.

## Madagascar Periwinkle 143
*Catharanthus roseus* (L.) G. Don

**Dogbane family** Apocynaceae

The genus *Catharanthus* comprises five species, four of which are native in Madagascar and one in India. The most widespread species, *C. roseus*, comes from Madagascar but is found nowadays as a naturalised plant in all tropical regions. It is semi-woody and reaches a height of 60cm. The round, hairless stem is usually branched and bears stalked leaves in a spiral arrangement. The leaf-blade is lanceolate and bluntly pointed. On the upper side it is dark green with a pale midrib and underneath it is light green and more or less downy. As a rule the flower-stalks are shorter than the leaf-stalks so that the beautiful pink flowers appear almost sessile. The calyx-lobes are hairy, narrow and pointed. The corolla too is softly hairy and has a purple throat and wide-spreading lobes. Besides the typical pink form there are cultivated varieties which are pure white or white with a pink or yellow throat. The plant is used in all kinds of ways for medicinal purposes. In Jamaica, for example, it is used to treat diabetes.

## Busy Lizzie 144
*Impatiens walleriana* Hook. f.
Syn. *I. holstii* Engl. & Warb.

**Balsam family** Balsaminaceae

The Busy Lizzie is popular everywhere as a house-plant and as an annual for summer bedding, and it can be propagated vegetatively without any difficulty. Its native country is the mountains of tropical Africa and Zanzibar. In those regions it produces branching, semi-shrubby growth to a height of 30–60cm. The stems are smooth, succulent and almost transparent, and the leaves are alternate at the base but in whorls in the upper part. The leaf-blades are elliptic to lanceolate, pointed at the apex, toothed along the edge, and they narrow into the stalk. The bracts at the base of the flowers are small and membranous. The corolla is up to 4cm across the wide-spreading lobes, and almost symmetrical but for one of the sepals which is prolonged backwards into a long slender spur. The colour of the corolla is usually crimson, but may also be scarlet, orange-red, pink or white.

## Glory Lily 145
*Gloriosa superba* L.

**Lily family** Liliaceae

The Glory Lily is native in tropical Asia and Africa and is a plant of outstanding beauty. In its natural habitat it climbs about the undergrowth of the forests, attaching itself by means of spiral tendrils at the end of its leaves. It has a tuberous underground rootstock from which arise the frequently branched stems, bearing leaves in pairs or in whorls of three. The flowers are produced in the axils of the upper leaves. In the bud stage they are nodding, but the six perianth-segments forming the corolla quickly become erect as the flower opens. The segments are wavy-edged, bright yellow at the base and fiery red at the top. This colouring shows the flowers in the intermediate stage for when young they are entirely yellow and when old completely red. The long stamens spread out from the centre of the flower and curve upwards, forming pivots for the narrow, delicately balanced anthers. The long, slender style extends horizontally from the top of the three-celled ovary and contributes to the unusually bizarre colouring and form of the flower. The ovary develops into a capsule containing bright red seeds. The name of the genus was derived from the Latin word 'gloriosus' meaning 'full of glory', and the genus itself contains five other species which also originate in Asia and Africa. All the species are very similar in having perianth-segments which always point upwards, and which remain attached to the fruit even when they have withered. Nowadays *Gloriosa* is frequently offered in flower-shops as cut-flowers.

## Golden Angel's Trumpet 146

*Brugmansia aurea* Lagerheim

**Nightshade family** Solanaceae

Angel's Trumpets are related to the Thorn Apple and have sometimes been included in the genus *Datura*. They are a systematically complex group of plants. There are about 20 different species which can be divided into two sections, woody and herbaceous. The species illustrated is a tree or tall shrub from Peru which may reach a height of 4m. The ovate-lanceolate, pointed, stalked leaves are sparsely covered with soft hairs on both sides and are arranged in groups. The lower ones are usually slightly toothed or lobed but the upper ones are entire. The conspicuous flowers are produced at the ends of the stems and are up to 30cm long. They have a long, slender corolla-tube which opens out into five reflexed lobes. The inflated calyx-tube is five-angled and has two or more large teeth on the margin. Like many members of the Solanaceae, the plant is poisonous. It is also grown in regions outside the tropics which are sufficiently warm.

## White Angel's Trumpet 147

*Brugmansia candida* Pers.
Syn. *Datura candida* (Pers.) Safford

**Nightshade family** Solanaceae

The White Angel's Trumpet is also a shrubby plant and closely related to *B. arborea* which is tree-like in form. It differs from this species in having a calyx which remains attached to the prickly capsule instead of separating from it when the fruit is ripe. The trumpet-shaped flowers of *B. candida* also reach a length of up to 30cm, but are distinguished by the absence of notches between the slender-tipped corolla-lobes. The ovary, formed from two united carpels, develops into a capsule containing numerous seeds. The natural distribution of this species is thought to be in the mountains of Peru and Chile. Related species are very similar in appearance. They are frequently cultivated and have probably hybridised. There is a preference for pink-flowered varieties. Many species and cultivated forms produce a pleasant scent during the evening and night. All parts of the plant are poisonous and have a narcotic effect.

## Aechmea

148

*Aechmea fasciata* (Lindl.) Baker

**Pineapple family** Bromeliaceae

The genus *Aechmea* comprises some 150 species which originate in Central and South America. The generic name is derived from the Greek word 'aichme', meaning 'point of a spear', and refers to the bracts of the inflorescence which are often sharply pointed. The plant is an epiphyte, and has 10–20 leaves arranged in a tubular rosette growing from a very short stem. The leaves are up to 50cm long and 6–8cm broad with a rounded apex tipped by a sharp spine, and edged with other dark, closely set spines. They are banded or marbled with silvery scales which help members of the pineapple family to take up water and any nutrients present. In some plants the whole leaf is covered with scales so that it appears white-felted, and in that case the roots have a less important role to play in the acquisition of nutrients. There are often interesting relationships between epiphytic Bromeliaceae and animals. For example, water collects in the leaf-bases of *A. fasciata* and tree-frogs find this a suitable place in which to spawn. The flower-stalk is erect, more than 30cm high, and covered with white woolly hairs. At the top is the dense pyramidal group of flowers surrounded by showy pink bracts toothed along the margin. In the axils of these bracts are the violet-coloured flowers with deeply fringed scales at the base of the petals. The inferior ovary develops into a berry. This species is native in Brazil but because of the increased interest in house-plants it is now one of the most popular and frequently grown representatives of the pineapple family outside tropical regions.

## Guzmania

149

*Guzmania lingulata* (L.) Mez var. *cardinalis* André

**Pineapple family** Bromeliaceae

The genus *Guzmania* comprises 120 species with a natural distribution in Central and South America and the Antilles, one species occurring wild in the southern United States. It was named in honour of the Spanish apothecary A. Guzman, who was also a naturalist and collector. The plant illustrated comes from the Antilles and is an epiphyte. Its stout leaves are up to 45cm in length and widen out at the base. They are green on the upper side and have a few reddish brown lines beneath, and in var. *cardinalis* are even broader at the base. The inflorescence, 40–50cm in height, grows out of the rosette of leaves and the outer bright red bracts are arranged in a spiral at the top. The inner bracts are yellowish and often almost transparent, and in the axils of these smaller bracts are the flowers, densely crowded together. All six perianth-segments are milky white. The three outer segments are erect and united up to halfway into a tube, but the three inner are connected only at their tips so that they form a bell-shaped structure with vertical slits. The stamens are attached for most of their length to the inner segments and have only their tips free. The three-celled ovary, with its slender, thread-like style, develops into a triangular capsule containing many seeds. The brilliant red bracts of this variety form a striking contrast to the green leaves. Like all members of the pineapple family, the bracts only acquire their red colouring as the plant comes into flower.

## Nidularium 150

*Nidularium* spp.

**Pineapple family** Bromeliaceae

The genus *Nidularium* comprises 30 epiphytic species native in Brazil, and the name is derived from the Latin word 'nidulum' which means 'little nest'. This name was chosen because the leaves are arranged in a rosette which encloses and protects the flowers just as a nest gives shelter to its occupants. The leaves are linear to lanceolate in shape and have sharp teeth along the edges. Brightly coloured outer bracts surround the dense group of flowers, and each flower arises in the axil of a small, membranous inner bract. The sepals are oblong, sometimes narrowly so, and the petals are united at the base or even further up, forming a tube. They end in a blunt tip, which either points upwards or is turned slightly to one side. In general the flower appears greenish at the base and white at the apex. The ovary is three-celled, and, where the cells meet, each cell contains at maturity a row of seeds. The genera *Nidularium* and *Neoregelia* are very closely related so that the number of species in each can vary according to the authority concerned. Both of these genera, and the genus *Canistrum* too, contain epiphytic plants of particular interest and charm.

## Neoregelia 151

*Neoregelia carolinae* (Mez) L.B. Smith var. *tricolor* L.B. Smith

**Pineapple family** Bromeliaceae

The genus *Neoregelia* comprises 45 species which originate in Brazil, Guayana and Peru. By far the majority come from Brazil. The name of the genus is derived from 'neo' meaning 'new' and from the name of the botanist E.A. von Regel, who lived from 1815–1892 and was director of the botanic garden in St Petersburg, now Leningrad. The plants in this genus are all epiphytic and have dense rosettes of leaves. The leaves have sharply toothed margins and are usually green though there are a number of cultivated varieties with the leaves striped. The upper leaves are notable for the brilliance of their colours which may be violet, blue, white or red. The inflorescence nestles in the cup formed by the smaller leaves at the top, and consists of a compact group of flowers each in the axil of a bract. In this genus the three sepals stand erect and are united to form a tube round the three petals which are also joined together with only their tips free. The three-celled ovary develops into a berry containing numerous small ovoid seeds. The cup formed by the upper leaves often fills with water, so that the flowers or fruits are no longer visible.

# Orchids

Orchids hold a special place amongst tropical plants and they have developed such a variety of form that it overwhelms the imagination. The number of species described totals 27,000 making the Orchidaceae one of the largest families in the plant kingdom. To this number can be added more than 50,000 cultivated varieties. Orchids are particularly abundant in the tropics and are often regarded as characteristic of tropical vegetation in the same way as palms. But orchids can be found outside the tropics, even in Greenland!

Orchids can be divided into two groups according to where they grow. Those that live on the ground are called terrestrial, and those that have their home in trees or on rocks are described as epiphytic. In the tropics the latter group exhibit an extraordinarily wide range of form, colonising not only the continuously damp areas such as rain-forests but also those places which have alternately wet and dry seasons like the savannah and steppe regions. In dry areas they adapt to the shortage of water in various ways. Some produce pseudobulbs (bulb-like enlargements of the stem) in which they can store water, others have fleshy, cylindrical leaves which reduce evaporation.

Tropical epiphytic orchids anchor themselves to the supporting branches by means of roots which have a special outer layer of tissue (velamen), enabling them to take up any water and dissolved nutrients in their vicinity. In extreme cases the plant may consist almost entirely of roots, pressed against the bark of the tree or spreading out into the air. These contain a considerable quantity of chlorophyll, and so, by the process of photosynthesis, the necessary organic substances can be produced.

One of the strangest facts about the life of orchids is that, after a flower has been pollinated, numerous seeds are produced (up to four million in the case of some tropical species) which are without any reserves of food. Consequently they are dependent for germination and further development on the presence of fungi (mycorrhiza), which provide them with nutrients while they are young or throughout their life, the degree of dependence varying according to the species in question. Some terrestrial orchids have been discovered that grow entirely underground and even flower below the surface.

The variety shown by orchid flowers is inexhaustible, but in spite of their varied appearance, they all conform to the basic plan of three sepals and three petals. One of the petals is usually conspicuously different from the others. This is called the lip or labellum. The structure of orchid flowers is intimately connected with the insects which pollinate them. For example, there are tropical orchids in the American genus *Oncidium* with flowers that resemble female bees (genus *Euglossa*). The flowers move in the wind and attract the attention of the male bees which then visit them bringing with them pollen from other flowers of the same kind. In the case of some species of *Cattleya*, another American genus, it is known that the flowers are visited by bees only during a period of 15 minutes before dawn. It is interesting to note that there are also some very unusual species of orchids in South Africa and Australia.

## Sukhakul's Lady's-slipper

**152**

*Paphiopedilum sukhakulii* Seng. & Schos.

**Orchid family** Orchidaceae

The name *Paphiopedilum* is derived from Paphia (Paphos was the birthplace of the goddess Venus) and 'pedilon' meaning 'slipper', and refers to the prominent, slipper-shaped lip of the flower. The genus comprises 65 species distributed throughout S.E. Asia, from India, across the Malayan archipelago and the Spice Islands, to New Guinea. The species illustrated is named after the Thai orchid collector Sukhakul and was not known until 1964, which shows that there are still plants in the world to be discovered and described. Like all species of *Paphiopedilum* this plant has a compact leaf-rosette consisting of four to seven narrowly elliptic leaves, up to 25cm long and 4–7cm broad. The dark brown flower-stalk is usually 25cm high and bears a single flower, 10–12cm across. The three outer perianth-segments or sepals are very different from the inner ones. The two lower sepals are joined together (a feature of the genus), directed downwards and scarcely visible behind the large lip. The third sepal points vertically upwards and has a background colour of white or yellowish green on which are a number of green stripes running along its length. Two of the inner perianth-segments or petals spread out horizontally sidewards. They usually have five, longitudinal, dark green stripes on a pale green background and dark coloured hairs along the margins. These petals are covered with reddish purple dots. The third petal, modified to form a lip, is relatively narrow, purplish brown near the aperture and shading to greenish at the toe. At the centre of the flower is the sterile stamen shaped like a convex disc.

## Ivory Lady's-slipper — 153

*Paphiopedilum bellatulum* (Rchb. f.) Pfitz.
Syn. *P. concolor* (Batem.) Pfitz.

**Orchid family** Orchidaceae

*P. bellatulum* is native in Thailand, Burma and Tonkin and grows there in humus-filled holes in limestone rocks. the Latin word 'bellatulum' means 'neat' or 'pretty' and the synonym 'concolor' refers to the uniform colouring of the flower. The plant has only about four leaves, marbled dark and greyish green on the upper surface and rich reddish purple beneath. They are broadly tongue-shaped, 10–15cm long and 4cm broad. The flower-stalk grows to about 10cm high and is covered with reddish hairs. It bears one or two flowers 7–8cm across. The sepals and lateral petals appear rounded at the apex, but the petals in fact end in a blunt point which is slightly bent back. In this species the lip is flattened at the sides and less striking than many others of the genus. The background colour of the flower is white or pale yellow to ochre, and the markings are in the form of small purple dots.

## Moth Orchid — 154

*Phalaenopsis amabilis* Bl.

**Orchid family** Orchidaceae

The genus *Phalaenopsis* comprises 70 species with a natural distribution extending from the Indomalayan archipelago to Formosa, N.E. Australia and New Guinea. The name of the genus is derived from the Greek word 'phalaina', meaning 'moth' and refers to the shape of the large and very beautiful flowers. Like all members of this genus the species *amabilis* (Latin for 'lovely') is an epiphytic orchid. It has a short stem with obovate or oblong leaves up to 30cm long and 12cm broad but no pseudobulbs. The slender, arching flowering-stem grows up to 80cm in length, bearing as many as 20 stately white flowers, 8–10cm across. The three sepals are broadly lanceolate and have blunt tips, while the lateral petals are broadly lozenge-shaped to circular and distinctly narrowed at the base. The lip is formed in a curious way. There are two lateral lobes which are striped red at the base and a diamond or tongue-shaped middle lobe with two long curled appendages at the end. The side lobes are curved to the same extent as the column (gynostemium) at the centre of the flower. On the lip is a raised, saddle-shaped structure, golden yellow and dotted with red, which is called the callus. The slender, semi-cylindrical column has a distinct foot and bears at its apex an anther with two round, grooved, waxy pollen-masses or pollinia. This species is native in Indonesia and New Guinea and includes a number of especially beautiful varieties, including var. *rimestadiana* which has pure white flowers with a tinge of yellow at the base of the lip.

## Dendrobium 155

*Dendrobium phalaenopsis* Fitzg.

**Orchid family** Orchidaceae

The genus *Dendrobium* comprises 1500 species native in tropical Asia from Ceylon to the Pacific islands of Samoa and Tonga and northwards to Japan. The name is derived from the Greek words 'dendron' (tree) and 'bioein' (to live). This species, so named from the resemblance of its flowers to those in the genus *Phalaenopsis*, has slender, cylindrical pseudobulbs up to 70cm long with oblong-lanceolate, dark green leaves that remain for many years. At the end of the pseudobulb is the slightly curved flowering-stem, up to 50cm in length, bearing 3–15 attractive flowers, 8cm across and variable in colour. The three sepals are broadly lanceolate, pointed, and pink to cherry red. The two lateral petals are more than twice as broad and usually of a darker hue. The lip is dark purple towards the base but its tongue-shaped middle lobe is the same colour as the unmodified petals. All the perianth-segments are more or less distinctly net-veined, and the bases of the two lateral sepals form a short spur behind the lip. The column of the flower is short and at its apex is the anther with four waxy, compressed pollinia, arranged in pairs. The home of this very popular orchid is in Queensland, Australia, and on the islands of Timor and New Guinea. It is a floriferous species, often producing flowers several times in the course of a year.

## Miltonia 156

*Miltonia* hybrid

**Orchid family** Orchidaceae

The genus *Miltonia* comprises 20 species with a natural distribution extending from Costa Rica to Ecuador and with the main centres in Brazil and Colombia. It was named after Viscount Milton, afterwards Earl Fitzwilliam, an orchid enthusiast. All members of the genus *Miltonia* are epiphytes and have short but well-formed pseudobulbs with one to three thin leaves. Erect flowering-stems, bearing one to several flowers, arise from the base of the younger pseudobulbs. The flowers of this genus can be distinguished by their flat appearance and by the disproportionately large and almost undivided lip, which has short outgrowths at the base. The sepals and petals are almost identical. The column is short and has two small auricles or wings at the top or in front. The anther has two pollinia on a short, triangular stalk. This genus contains numerous hybrids, often of great beauty, and many of the cultivated varieties are named after towns or other geographical localities.

## Masdevallia

157

*Masdevallia tovarensis* Rchb. f.

**Orchid family** Orchidaceae

The genus *Masdevallia* is named after the Spanish doctor and botanist Masdevall who died in 1801, and the name of the species is derived from the Tovar region of Venezuela where the plant occurs at an altitude of about 2000m. There are 300 species in the genus, all epiphytic, and native in the cool, wooded, mountainous regions of South America. The elliptic to spathulate leaves, 12–15cm long and notched at the tip, arise from the creeping rootstock or rhizome. *M. tovarensis* produces a triangular flowering-stem, 15–18cm high, which bears delicate white flowers for two or three years. Sometimes two to four flowers in the inflorescence are open at the same time. The three outer perianth-segments (sepals) are joined together and, as in all species of *Masdevallia*, form the showy part of the flower. The distance between the tip of the sepal pointing upwards and those of the two downward-pointing sepals amounts to 8–10cm. The needle-like tips of the sepals are tinged pale yellow. The inner perianth-segments (petals) are greatly reduced in size. The same is true for the lip, which is attached to the foot of the column and is also white and insignificant. The anther is one-celled and has two waxy, unstalked pollinia. The flowers are remarkable for the translucent appearance of their sepals and the snow-white lines running along them, and also for their rather bizarre shape. The long, slender tips of the sepals are often referred to as 'tails' and in some species can be as much as 20cm in length. There is also a wide range of colours within the genus, from pure white, through various shades of yellow to brilliant red and black-purple.

## Angraecum

158

*Angraecum eburneum* Bory

**Orchid family** Orchidaceae

The genus *Angraecum* comprises 200 species with a natural distribution in tropical Africa. 120 of these are found on the island of Madagascar, and one species is native in Ceylon. The name *Angraecum* is derived from a Malay word meaning 'air plants', a reference to the epiphytic nature of the orchids. The name of the species is taken from the Latin word for the colour 'ivory-white'. *A. eburneum* is one of the Madagascan species and produces a stout, leafy stem up to 1m in height, with leaves that are folded lengthwise and point upwards at an angle. They are up to 50cm long and 5cm broad and have two unequal lobes at the apex. The inflorescence, sometimes erect or leaning to one side, consists of up to 15 fragrant flowers, set closely together at the top of the flowering-stem and often facing in the same direction. The flowers are 8–10cm across and are remarkable in having the lip or labellum at the top. This is, in fact, the natural position, but in many orchids the individual flower-stalks twist round, bringing the lip to the bottom of the flower. The lip of *A. eburneum* is tongue or kidney-shaped and long-pointed, ivory-white at the edge shading to green towards the base, and with a fleshy outgrowth along the middle. This species is distinguished not only by the position of the lip but also by a slender spur up to 8cm long. The other five perianth-segments differ little from each other. They are all narrowly tongue-shaped, pointed, pale green and glossy. The column of the flower is very short and the pollen is collected together into two globular masses of a waxy consistency.

Mention must be made of another species of Madagascan origin, *A. sesquipedale*, which has a spur up to 45cm in length ('sesquipedale' is Latin for '1½ft'). This plant is visited by a kind of moth with such a long proboscis (22.5cm) that it can reach the nectar which collects at the tip of the spur. The plant was first found in the early 1820s and later in the century Charles Darwin suspected the existence of a suitably long-tongued insect visitor, but it was not until 1903 that the moth was finally discovered.

## Coelogyne **159**

*Coelogyne cristata* Lindl.

**Orchid family** Orchidaceae

The genus *Coelogyne* comprises 150 species with a distribution extending from the Himalaya, through southern China to New Guinea, the Fiji Islands and Samoa. The generic name is derived from the Greek words 'koilos' (hollow) and 'gyne' (female) and refers to the concave stigma. It is an epiphytic species native in the Himalaya at an altitude of 1600–2300m, and has been cultivated outside tropical regions for hundreds of years. The rootstock of the plant bears a compact group of almost spherical, green pseudobulbs which soon become wrinkled. Each pseudobulb has two lanceolate pointed leaves up to 30cm long. The pendulous inflorescence arises from the base of the most recent pseudobulb, and has five to nine pleasantly scented flowers, 9–10cm across, and looking as though they were made of fine china. The narrowly elliptic, snow-white perianth-segments are wavy along the margins and slightly recurved at the tip. The three-lobed lip (labellum) is directed forwards and has five orange-yellow crests running along the centre. It is because of these crests that the name '*cristata*' was chosen for the species. The rounded side-lobes and the middle lobe of the lip are all slightly recurved at the edges. The column is slender and bears an anther with four pollinia.

## Odontoglossum **160**

*Odontoglossum bictoniense* Lindl.

**Orchid family** Orchidaceae

The genus *Odontoglossum* comprises some 300 species distributed from Mexico through Central America to Bolivia and Brazil and with a centre of concentration in the Andes. The name is derived from the Greek words for 'tooth' and 'tongue', and refers to the shape of the lip and the tooth-like outgrowths at its base. The species illustrated is an epiphyte, native in Mexico and Guatemala southwards to Costa Rica. It forms ovoid, somewhat flattened pseudobulbs, 12–15cm long, at the end of which are the narrowly elliptic, pointed, soft leaves, up to 50cm in length and pale green in colour. The erect flowering-stem, up to 1m long, arises from the sheaths at the base of the pseudobulb and usually bears 8–12 flowers with slender, inferior ovaries in the axils of scaly bracts. The flowers are 4–5cm across and the three outer and two of the inner perianth-segments are lanceolate in shape. Their background colour is reddish brown, and this is curiously patterned with bright yellow markings which often resemble more or less complete transverse bands. The pale pink lip is heart-shaped, slightly curled at the edge, and tapers to a short point. The column (gynostemium), which results from the union of the ovary with the single stamen, appears as a slightly curved structure at the centre of the flower. This species is found growing wild in the mild, damp mountain forests of Central America and is widely cultivated in cool greenhouses.

## Oncidium

161

*Oncidium nubigenum* Lindl.

**Orchid family** Orchidaceae

The genus *Oncidium* comprises 750 species with a natural distribution extending from Florida through Mexico and Central America to Paraguay and Argentina. Its name is derived from the Greek word 'onkidion', a 'swelling', and refers to the ridged or warty appearance of the lip. All the species are epiphytes and their inflorescences can be several metres in length. *O. nubigenum* produces from the axis of the stem a number of ovoid, somewhat flattened pseudobulbs, 4–6cm long, each with one or two strap-shaped leaves up to 15cm long. The arching inflorescence, which may reach a length of 50cm, arises from the base of the pseudobulb, and bears 6–12 flowers, each 3cm across. The sepals and lateral petals are chocolate-brown or dark greenish brown to dark purple. The dorsal sepal is oblong-ovate, pointed, and slightly wavy along the edges, like the two lateral petals. The other two sepals are united into a single perianth-segment with two tips, and are hidden behind the lip which is vaguely fiddle-shaped. The lip has two small round lateral lobes at its base and two larger, almost kidney-shaped, wavy-edged lobes at the front. The whole lip is coloured white or pink with prominent crimson spots. Between the two lateral lobes is a bright golden yellow outgrowth or callus. The column is short and compressed, and the anther is concave. Most species of *Oncidium* can be recognised by their pale or deep yellowish brown flowers with a fiddle-shaped lip.

## Cattleya

162

*Cattleya granulosa* Lindl.

**Orchid family** Orchidaceae

The genus *Cattleya* comprises 65 species, most of them native in Brazil and the tropical parts of the Andes, but some also in the tropical forests of Central America. The genus was named after the English botanist and gardener William Cattley and the species *C. granulosa* originates in Guatemala and Brazil. Like all members of the genus it is an epiphytic plant with a stout rootstock from which arise slender, stem-like pseudobulbs 40–60cm high. Each pseudobulb bears a pair of leathery leaves about 15cm long, oblong-lanceolate in shape and bluntly pointed. The short flowering-stem produces five to eight glossy flowers, 8–12cm in diameter. They are strongly scented and last for a long time. The three sepals are narrower than the lateral petals. The dorsal sepal stands erect, while the other sepals are curved and point downwards. The two lateral petals are lanceolate with a blunt tip and have wavy margins. These five perianth-segments are olive-green with dark red dots. The lip is distinctly three-lobed. The two lateral lobes, whitish yellow on the outside and yellowish to pink within, curve closely round the column. The middle lobe is much larger and is fan-shaped. Its central portion is rough from numerous small outgrowths, and its margin is finely fringed. It is white at the front with a sprinkling of pink to red dots, and yellow at the base with red lines. The species acquired its name because of the granular area on the lip. The four pollen-masses attached to the column (gynostemium) are in two pairs and have a waxy consistency.

The genus *Cattleya* contains a large number of highly decorative species, which, because of their beauty and splendour, are amongst the best known and most popular orchids. Members of this genus have been crossed with those in other genera, e.g. *Laelia*, *Brassavola* and *Sophronitis*, so that bigeneric and trigeneric orchids have come into existence. A well-known example of the latter is *Brassolaeliocattleya*.

# Water Plants

## Royal Water-lily

163

*Victoria amazonica* (Poeppig) Sowerby
Syn. *V. regia* Lindl.

**Water-lily family** Nymphaeaceae

The genus *Victoria* was named in honour of Queen Victoria who came to the throne in 1837. It has only two species, both of which are native in South America. *V. amazonica* is found in the quiet waters of the Amazon as well as in Bolivia and Guayana from 4° north to 15° south of the equator. The more or less circular floating leaves are attached by thick, strong, rope-like stalks to the rootstock, which is anchored in the muddy river-bed. The leaves are normally 2–3m in diameter but explorers have reported even larger leaves, up to 4m across, in favourable conditions within the natural distribution. The edges of the leaves are turned up vertically, like those of flan dishes, but at the point nearest to the attachment of the stalk, and also directly opposite, there are gaps so that, during heavy tropical rainstorms, the water can quickly drain from the surface of the leaves. The underside of the leaf and the stalk have sharp prickles, and when the leaf unfolds it resembles a small, round, prickly boat. The upper surface of the leaf is glossy green and under a lens has the appearance of a fine sieve. The underside is a coppery red and has a network of protruding ribs and veins which increase both the buoyancy and the stability of the leaf. The leaf can support a weight of 40–75kg if the load is evenly distributed.

In contrast to the leaves, which float on the surface of the water, the flowers rise a little above it while they are fully open, but sink down again as they wither so that the fruits ripen in the water. The flowers are very fragrant and have a diameter of 25–40cm. The perianth-segments are not clearly distinguishable into sepals and petals, but the four outer segments are coloured green on the outside and enclose the flower in the bud stage. The numerous inner segments are spirally arranged and gradually change to stamens as they approach the centre of the flower. While the innermost stamens have clearly defined anthers on a narrow stalk, those further from the centre become increasingly petal-like, and have broader stalks and smaller anthers. The view that petals evolved from stamens is supported here in no uncertain way. The petals, like all the other parts, contain air-passages which contribute to the buoyancy of the plant.

The white flowers of the Royal Water-lily open for two nights. They open at dusk on one day and close the following morning. In the afternoon they open a second time and close finally the next day, sinking below the surface of the water. During the flowering-time their colour changes from white to pink and then to red. This change of colour is connected with the rapid changes in metabolism which occur in tropical conditions. The intense katabolic processes (respiration) produce so much heat that the temperature inside the flower can rise to 10°C above its surroundings. Pollination is carried out by beetles. The prickly fruits, larger than a fist, contain numerous dark olive-green seeds, as big as peas, which later turn black and are eaten like maize by the Indians.

The discovery, description and scientific naming of the plant have been erratic. It was first discovered in 1801 by the German botanist Thaddaeus Haenke, who found it on the Rio Marmora, a tributary of the Amazon. However, this fact was not made public until some years after his death in the Philippines in 1817. It was subsequently re-discovered on a number of occasions, and named and described by D'Orbigny, but later its name was altered. The first seeds to germinate in Europe were in Kew Gardens, near London, in 1846, but these plants never reached the flowering-stage. It was only in 1849, in the glasshouses at Chatsworth, the home of the Duke of Devonshire, that plants were first brought into flower.

The other species related to *V. amazonica* is *V. cruciata* which is found growing wild in northern Argentina and Paraguay. Its leaves are somewhat smaller, green on the underside with reddish ribs. This species, from the edge of the tropical region, is further distinguished by the height of the upturned margin of the leaf. It is 12–18cm high, inflated, and an intense brownish red. In *V. amazonica*, by contrast, it is only 4–6cm high. A hybrid between the two species is in cultivation.

## Water-lily

**164, 165, 166, 167**

*Nymphaea* hybrids

**Water-lily family** Nymphaeaceae

The genus *Nymphaea* comprises 40 species distributed throughout almost the whole world including the tropics. Most of them have brightly coloured flowers, especially the hybrids, whose parentage can often no longer be ascertained. As a general rule it may be said that the more colourful the flower the more complex is its genetic history. The name of the genus is derived from 'nymphaia', for, according to Greek legend, the flower arose from a nymph, who had died of jealousy. On the other hand, according to Pliny, it arose from a nymph who died from unrequited love for Hercules.

All species of *Nymphaea* form vertical or horizontal, creeping rootstocks in the mud and can be uniform in thickness or tuberous in shape. The regularly formed rhizomes can be as thick as a man's arm and are patterned on the surface with the scars left by fallen leaf-stalks. The long stalks arising from the rootstocks end in circular, oval or heart-shaped blades which float on the surface of the water. The leaf-blades are shining green on the upper side, while the underside is differently coloured, often bluish green or tinged with red. The strikingly lovely flowers either float on the water-surface or rise above it. They consist of a large number of spirally arranged perianth-segments, which, like the Royal Water-lily, show all the intermediate stages between stamens and petals. Again, the four outer perianth-segments can be regarded as sepals. The numerous carpels are sunk in the receptacle, giving the appearance of a single fruit. The fruit ripens under water and breaks up at maturity. The rhizomes of many species are edible.

## Lotus

**168**

*Nelumbo nucifera* Gaertner

**Water-lily family** Nymphaeaceae

Like the genus *Victoria*, the genus *Nelumbo* comprises only two species, and its name is derived from the Cingalese word for the Lotus flower. The natural distribution of *N. nucifera* extends from Japan to N.E. Australia and across to the Caspian Sea. In eastern Asia the plant has long been cultivated and often becomes naturalised.

The creeping rootstock anchors itself in the mud and sends up 1–2m long stalks ending in circular, slightly funnel-shaped, bluish green blades, 30–60cm in diameter. The leaf-stalk is prickly and is attached to the centre of the shield-shaped blade. The upper side of the blade has a waxy surface, and drops of water roll about on it like globules of mercury. The flowers usually stand a little higher than the leaves. They are 18–35cm in diameter, and have a number of perianth-segments, normally pink but sometimes another colour such as yellow (see photo). The numerous stamens are typical of the flower. A remarkable feature is the receptacle in the form of an inverted cone in which the separate carpels are embedded. The round stigmas give the impression of holes so that the entire structure resembles the rose of a watering-can. After fertilisation the receptacle becomes enlarged and the nut-like fruits ripen inside. Like the rhizome these are edible.

The Lotus-flower is known as 'padma' in India and is venerated as a symbol of the Ganges. It has been spread westwards by man as far as Egypt, where it is also treated as a sacred plant. It has often been used as a motif in architectural decoration.

## Water Hyacinth **169**

*Eichhornia crassipes* (Mart.) Solms

**Pickerel-weed family** Pontederiaceae

The genus *Eichhornia* comprises six species with a natural distribution in tropical and subtropical America, and was named after the Prussian minister J.A.F. Eichhorn. *E. crassipes* is a water or swamp-plant with a short stem bearing numerous leaves arranged in a rosette. The leaf-stalks are distinctly swollen, which is why the species is called *crassipes*. The tissue inside the stalks contains air-cells, increasing the buoyancy of the plant, and the size of the inflated portion can vary according to the locality. The leaf-blade is smooth and pale green, and is broad-ovate to kidney-shaped. The inflorescence resembles a spike but the individual flowers are shortly stalked. The flower consists of six more or less equal perianth-segments, pale blue to pale purple in colour, and united only at the base. The uppermost segment stands upright, and is further distinguished by a network of darker veins and a bright yellow spot. The radial symmetry is also broken by the form of the stamens, which project from the centre of the flower and curl upwards at the end. The three-celled ovary develops into a capsule. The plant is remarkable for its numerous roots which are clothed with fibres spreading at right angles and have a comb-like appearance. The floating rosettes reproduce themselves by means of runners which arise in the axils of the leaves and form new rosettes at the ends.

The Water Hyacinth was introduced into India during the last century and, for want of any natural restrictions to limit its spread, it has quickly become a nuisance. The runners from one plant can cover an area of $100m^2$ in a few months, and, because of this rapid reproduction, canals can easily become blocked, and ponds made unusable for fish-farming. Herbicides have had little success so far in controlling infested areas, and attempts have been made to make use of the plant in various ways, e.g. as pig-food, compost, in paper-making and as a source of potash.

2 3

4 6 7 5

8 9 10

11

12

13

6 14 15

17

19

20

1

21

22

25 23 24

26

27

28

29

30

31

32

33

34

35 37 38 36

39

40 41

42

43

44

45

46

47

48

49

50 51 52

53

54

55

6 58 59 57

60

62

63

4

65

66

67

68

69

71

73

74

72

75

76

77

78

79

80

81

82

85 83 84

86

87

88

89

90

91 92

93

94

95

96

97

98

99

10

101

102

4 106 107 105

110 112 113 111

114

115

116

117

118

119

120

121

122

123

124

125

127

128

126

129

130

131

132

133

34

136

137

135

39

140

141

142

143

144

45

146

147

49

150

151

154 152 153

155 157 158 156

159

160

161

163

164

165

166

167

168

169

170

171

172

173

174

175

76

177

178

179

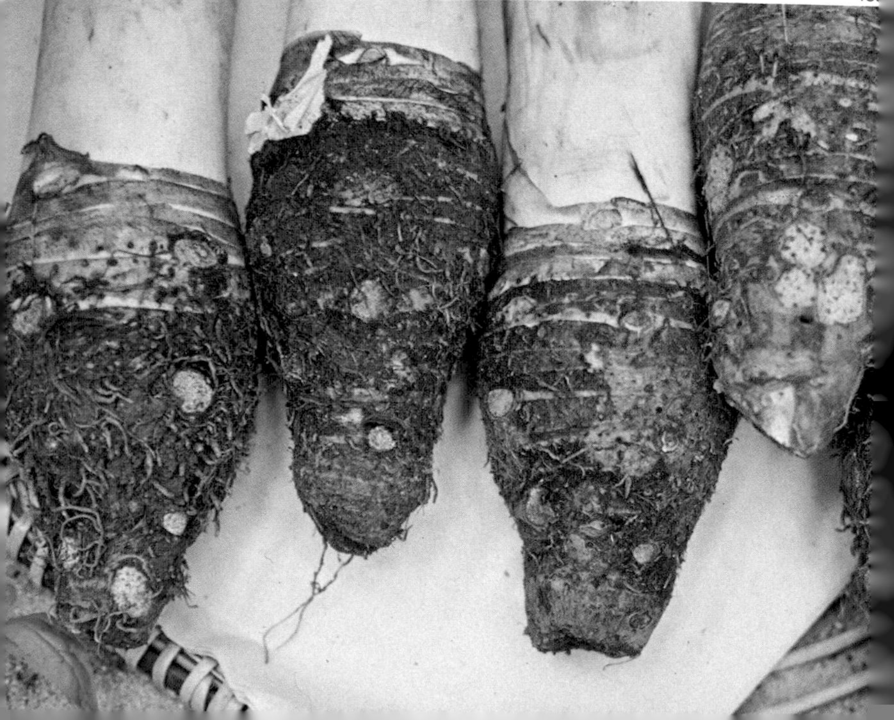

18(

83

181

182

184

185 186

187

188

189

190

191

192

193

194

195

196

197

199

200 201

203 204 206

209

207

208

21(

213 211 212 214

215

216

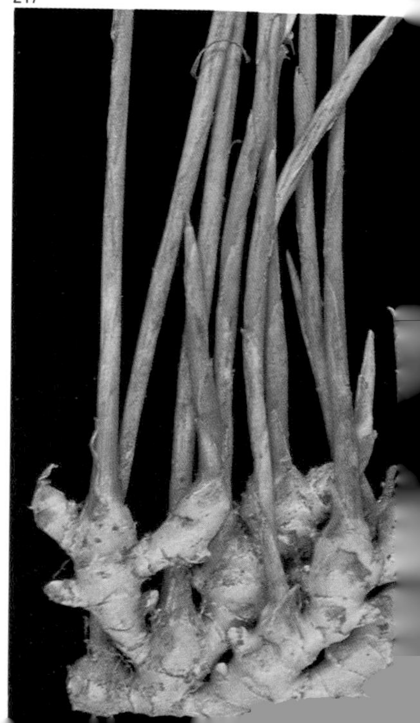

217

218

219

220

221

222

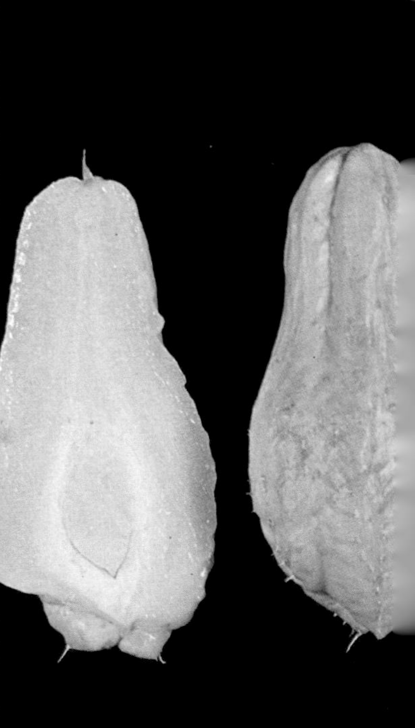

224

22

226

227

228

229
230
23

232

233

234

235

236

237

238

239

240

241

242

243

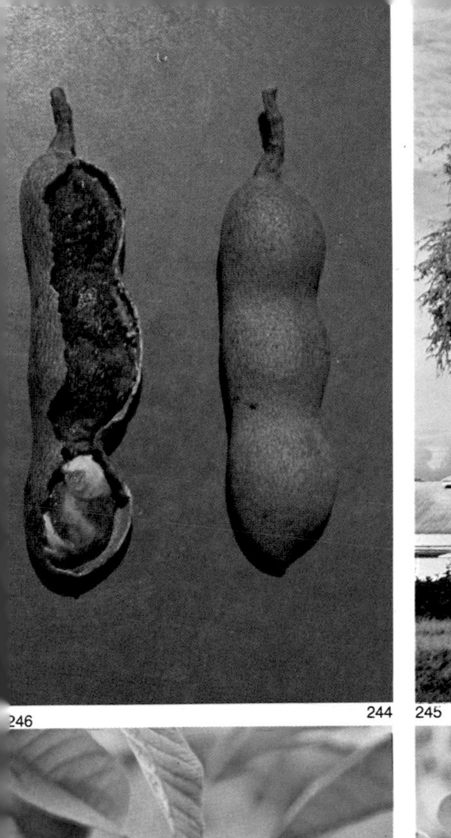

244

245

246

247

248

249

250

25

252

253

254

255

256

257

258

259

260 261

262

263

264

265

266

267

268

269

270

271

272

273

274

# Economic Plants

The number of tropical economic plants is very large and they have a very wide range. Their distribution, too, varies enormously. On the one hand there are plants which are widespread throughout the tropics, and on the other there are some that have only a local distribution. Only the first group fall within the scope of this book. Even today there are economic plants, such as those used by Indians on the upper Amazon and the Orinoco, whose use is still not fully understood.

Tropical economic plants can be grouped according to the purposes for which they are used. First there are the starch plants, which are the most important food sources for the inhabitants of tropical regions. Some of the best known are rice, maize and millet. Others, confined entirely to the tropics, are sweet potatoes, manioc (cassava), taro, yams, okumo and breadfruit. Yet others, of only local importance, are seeds of quinoa, and underground tubers of oca, ullucu, and species of *Tropaeolum*. Sugar cane, too, can be classed as a tropical source of carbohydrates.

Next comes a group of plants which supply protein, and often oil too. Important representatives of this group are beans, groundnuts, Brazil nuts, and akee fruit, also many palms such as the coconut palm. The fruits of a large number of palms are rich in protein and oil, and also contain carbohydrates, but are only suitable as food for animals. Many cultivated forms of economic plants are very decorative as well, and consequently may be regarded as ornamental plants. This is true not only for many of the palms but also for the breadfruit tree and many other tropical fruit trees.

Tropical fruits form a really extensive group, and these come from shrubs and climbing plants as well as from trees. Many are cultivated in all the tropical regions of the world, while others are grown only in S.E. Asia. Some of these never reach temperate zones, even in this era of air travel, because of their peculiar taste or smell, susceptibility to transport, or simply lack of value. Certain fruits, such as the banana, can be classified in more than one way, e.g. as a fruit and as a starch plant.

Finally there is a group which includes plants which are sources of spices, beverages, flavourings, food dyes and masticatories. Stimulants and medicinal plants are so abundant that only a few of these can be dealt with here. The same holds good for dye-plants and those where seeds or other parts of the plant are used for personal decoration. Dye-plants are used not only in the manufacture of textiles but also provide colourings for foodstuffs or even the human body. Mention must also be made of the inexhaustible stocks of tropical timbers. Many of the plants in these varied groups are only of local significance, but some of them contain valuable substances which cannot be produced synthetically and yet have great medicinal importance. As tropical economic plants can usually be evaluated in more than one way it is often difficult to refer them unequivocally to one particular category. In the following section of this book, an attempt has been made to place in the same group all those plants which are used primarily for the same purpose.

# Starch Plants

## Rice

**170, 171, 172**

*Oryza sativa* L.

**Grass family** Gramineae

Although rice is predominantly a tropical starch plant, its cultivation extends into the subtropical regions of the world. Rice-fields can be found from 45° north to 40° south of the equator. After wheat and maize it is the most widely grown cereal and, with a yield of four tons per acre, it is the most productive of them all.

The history of rice as a food-plant can be traced back to 6000 BC. Its cultivation in India and China presumably developed independently, and spread from these early centres of cultivation to Japan, Australia, South America and California, and also to West Africa, Egypt and Madagascar.

The plant reached the mediterranean region about AD 800 and is still grown today in Italy, Spain and southern France. Its native country is probably India, where it still occurs in the wild, but there are also wild plants in South America, Africa and Australia. Swamp or lowland rice is the kind most frequently cultivated, but hill or dry-land rice is also grown. A whole range of varieties is in existence, and also numerous cultivated forms which belong to genetically related groups in South America, Africa, India and S.E. Asia. They vary from erect plants only 30cm high to floating forms, suitable for areas of deep flooding, which reach a length of 7m.

Rice is a grass with the inflorescence in the form of a panicle. In areas of intensive cultivation it is either sown broadcast and later transplanted by hand or individual seeds are put into the ground by mechanical means. The seedlings grow quickly and are sometimes ready for transplanting after only 20 days. At first, the stem, which will eventually reach a height of 1.8m, has only lower leaves in the form of leaf-sheaths, but later the leaf-blades grow to 50cm in length, or even longer if they are in water. The panicles (Photo 172) grow up to 50cm long and bear three-flowered spikelets, only the terminal flower of which is bisexual.

The plant is very demanding in its requirements of temperature, nutrients and water. The soil must be rich in humus and nutrients. In the case of swamp rice (Photo 171) the plants stand in mud below the surface of the water and when being transplanted the soil must be flooded to a depth of 15–30cm (Photo 170). Part of the nitrogen requirements are met by blue algae which live in the water and bring about fixation of nitrogen in the vicinity of the roots. The temperature must be between 25° and 30°C. After the plants have flowered, the water-level is gradually lowered until, at harvest-time, the fields are dry. Upland rice can be grown up to an altitude of 1800m, but even this needs a high rainfall and it thrives best in areas of rain-forest which have been cleared by burning. Its demands of temperature are not so great and an average of 18°C is sufficient.

According to the variety concerned and local conditions, a period of three to nine months elapses between sowing and harvest. In some regions, therefore, three or even four crops can be harvested in a year. Harvesting is usually carried out by hand, using a sickle, but combine harvesters are used in the USA and Australia. In Asia, threshing is often done by water-buffalo, which are allowed to trample over the sheaves until the chaff has been separated, but even there threshing machines are being used in increasing numbers. The unhusked grain, as well as the growing crop, is known as 'paddy'.

The great economic significance of rice is well-known, and, particularly in the developing countries of Asia, it is the most important source of carbohydrates. More than 60% of the total world crop is grown in China, India and Indonesia, but, because of the rising population, home-grown supplies have now proved insufficient, and countries such as India, Indonesia and Vietnam have had to import extra quantities from the USA. The yield varies enormously, according to the standard of agricultural development, from 0.65 tons per hectare in Senegal to 7.9 tons in Australia.

## Sweet Potato **173, 174, 175**

*Ipomoea batatas* (L.) Lam.

**Convolulus family** Convolvulaceae

The sweet potato forms a root-tuber like the manioc or cassava plant. It is cultivated throughout the tropics and in parts of the subtropics and provides ample nourishment. Its tuberous root contains mainly starch but also so much sugar that it has a sweet taste. Its natural distribution is very probably in South America, since 'batata' is the Carib name for the plant, also from relics found in Peru it would appear that the plant was already being cultivated there in pre-Columbian times. It is also evident that the sweet potato arrived in England, via the Canary Islands, at the beginning of the 16th century, before the ordinary potato made its appearance there. The name 'batata' was subsequently altered to 'potato' and later transferred to the other vegetable. It is thought to have been introduced into the Philippines and Moluccas by the Spaniards.

In order to thrive, the sweet potato needs a frost-free climate, in fact a temperature of 10°C is too cold for it. The most desirable conditions are an annual average temperature of 26°–30°C and an annual rainfall of 850–900mm, but the wet spell should coincide with the plant's growth period and the weather should be dry for the plant to ripen properly. It can be cultivated up to 40° north and south of the equator, and up to an altitude of 2200m, as long as the average temperature does not fall below 22°C. If the soil is slightly acid and well drained, the tubers will ripen in four to five months.

The sweet potato is a perennial herbaceous plant but is grown as an annual and, like some other members of the Convolvulaceae, creeps about on the ground (Photo 173). The stems, several metres long, are produced from a root-tuber that becomes exhausted during their growth. The leaves are arranged alternately on the creeping stems and are very variable in shape. They are always distinctly stalked, but the leaf-blade may be heart-shaped, sometimes with pointed lobes, or deeply divided into broad or narrow segments, as shown in the drawings overleaf. In addition to the plants with purely green leaves there are varieties with leaves tinged purple beneath because of the formation of anthocyanin. Present-day plant breeding tries to select forms which have short stems and bushy growth. The nodes of the creeping stems produce roots which not only serve to take up nutrients but also store carbohydrates. In doing so they swell and form spindle-shaped tubers (Photo 175). These may weigh as much as 3kg, are pale red, yellowish brown or whitish in colour, and represent the sweet potato of commerce. When the tuber is cut, it exudes a milky juice which is not poisonous.

The sweet potato flowers as a 'short-day' plant in the tropics, developing an inflorescence of three or four funnel-shaped white or reddish flowers. The two-celled ovary develops into a capsule containing two to four very hard, black seeds. In cultivation the plant is propagated vegetatively by rooted sprout or stem-cuttings.

Yields vary considerably, according to the location and standard of cultivation, from 0.5 tons per hectare in Mauritania to 48 tons in Italy. The annual average for production throughout the world is about 16.8 tons per hectare. The quantity

**Production of Sweet Potatoes** (in millions of tons)

| *Country* | *1975* | *1978* | *1980* |
|---|---|---|---|
| China | 114.4 | 78.5 | 87.1 |
| Indonesia | 2.5 | 2.5 | 2.0 |
| South Korea | 2.0 | 1.6 | 1.1 |
| Brazil | 1.7 | 1.9 | 1.0 |
| India | 1.7 | 1.7 | 1.6 |
| Japan | 1.4 | 1.4 | 1.4 |
| Other countries | 11.9 | 12.2 | 13.1 |
| Total | 135.6 | 99.8 | 107.3 |

produced by individual countries can be seen from the above table. The nearer the place of cultivation is to the equator, the higher is the sugar content of the tuber. The tubers are eaten boiled or roasted, and when dried can be cut and ground into flour, starch and a kind of sago. In the West Indies and Latin America the tubers, rich in sugar, are allowed to ferment in order to produce alcoholic drinks called 'mobby' and 'mormoda'. On Easter Island, sweet potato tubers have for centuries been the islanders' staple food.

*Ipomoea batatas.* Various leaf-shapes of varieties grown in the Caribbean area.

## Cassava, Manioc, Tapioca

176

*Manihot esculenta* Crantz

**Spurge family** Euphorbiaceae

Cassava is one of the most important food-plants in humid tropical regions. Its homeland lies in the New World, although its exact place of origin is not known. Its natural distribution is likely to have been in Venezuela, Brazil and Central America, but the form cultivated nowadays is probably the result of hybridisation between different species. Extensive cultivation of cassava began in Brazil, Colombia and the West Indies, and has spread across the tropical regions of West Africa to India, Thailand and Indonesia.

Cultivation of the plant goes back to 2500–3000 BC. In the 16th century the Portuguese brought it to West Africa, where it quickly spread south of the equator, but was less enthusiastically received in East Africa. The plant first reached Ceylon in 1786 and Java in 1835.

The requirements for the formation of good root-tubers are a warm climate, with an average annual temperature of 20°C, and an annual rainfall of 500–5000mm with not too high humidity. The plant needs a great deal of light but its demands on nutrients in the soil are moderate. It will not tolerate waterlogged ground, therefore the soil must be deep and well-drained. Cassava is a herbaceous perennial which lasts for several years. Its stems grow up to 5m in height and become woody with age, bearing scars where the leaves have fallen off. The leaves are arranged alternately and have long stalks and digitate, five to nine-lobed blades. The shape of the leaves can vary considerably between individual cultivated varieties. The inflorescence is in the form of a panicle at the end of the stem, and consists of a few female flowers and, above them, numerous male flowers. There may be as many as 200 male flowers and only 20 female, but the latter open first so that cross-pollination is usually assured. The three-celled capsules spring open suddenly, flinging out the seeds.

At the base of the stem, the plant forms conical to spindle-shaped root-tubers, like the dahlia. These are 30–50cm or even 90cm long and 5–10cm in diameter, and weigh 2–5kg (illustrated p. 182). They are rich in carbohydrates but relatively poor in proteins. Like all members of the Euphorbiaceae the plant has a milky sap containing the poisonous cyanogenetic glycoside linamarin. Prussic acid is released from this by an enzyme present in the cell-tissue. However, boiling, steaming or roasting the peeled tubers is sufficient to destroy the dangerous linamarin. South American Indians are able to extract the starch, before the tubers are eaten, with a special little tool called a 'tipiti'. This is a tube, 20–30cm long, made from woven palm fibres, which can be shortened or lengthened by pressing and pulling. The raw cassava-roots are first pounded or grated into small pieces. Then the mash is put into the tipiti and kneaded so that the starch collects in the lower part of the tube.

In spite of its modest requirements, cassava is a tropical starch-plant with a high yield, but as its tubers cannot be stored for long and do not all ripen at the same time, it is not suitable for growing on a large-scale. It is much more often cultivated, in small plantations or in plots of land round huts, from stem-cuttings taken as required. If the plants remain there longer, they produce new young shoots. Harvesting takes place after 6–24 months when the leaves begin to change colour, 12 months being the average time. Sweet varieties ripen earlier than those that are more bitter. The tubers must be processed soon after harvesting, since they rapidly turn blue and begin to rot, losing 10–20 per cent of starch in only two months. Yields vary between 2.5 tons per hectare in Guinea and 26.5 tons in Barbados.

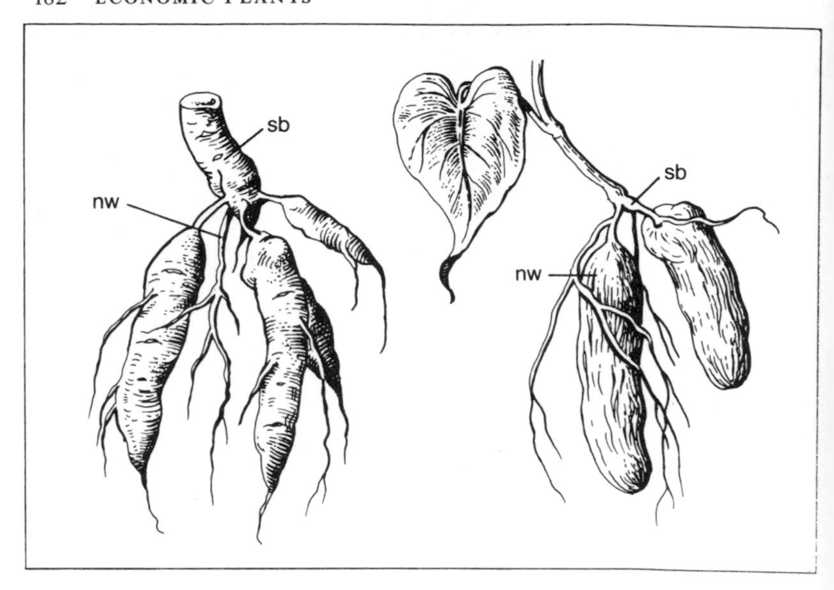


*Left:* Root-tubers of the cassava (*Manihot esculenta*) *Right:* Root-tubers of the yam (*Dioscorea* sp.). sb = base of the stem; nw = roots for absorbing nutrients and water.

In the tropical regions of cultivation the tubers are often boiled, mashed and eaten with a sauce or baked into flat, round cakes. A large number of them are ground into flour, either by hand or by machines in factories. In this case the tubers are peeled, washed, and grated to a white mass. They are then dried, first in the sun, and afterwards on hot plates. After grinding, a usable, non-toxic flour is obtained, that is known in the trade, especially in Brazil, as 'farinha'. In order to obtain starch for export the tubers are grated, and the grated mass then kneaded in water so as to remove the milky juice. The starch that remains is again washed and dried in the sun. It supplies the commercial product tapioca which can be also turned into sago.

## Potato Yam
*Dioscorea bulbifera* L.

**Yam family** Dioscoreaceae

The potato yam is distinguished by the aerial tubers which form on the twining stems. It is for this reason that it is also known as the 'air potato'. These tubers are produced in considerable numbers in the axils of the leaves and weigh 500–2000g. If the tuber is cut open the inside will often be found to be dark purple due to the anthocyanin content of the cells. In the case of the potato yam the aerial tubers can be used as food but those produced underground are unpalatable. The natural distribution of the plant can no longer be ascertained, but it is now widely cultivated, especially in Micronesia, Indonesia, India, Japan, northern Australia and Africa. Judging by its widespread distribution in Micronesia it is possible that this is its native area, although it is also said to occur wild in India and Africa.

Most species of *Dioscorea* are tropical plants and require an average annual temperature of 20°C, high humidity, regular rainfall and light, humus-rich soil. They are 'short-day' plants and need the equal balance of daylight and darkness in order to produce tubers. Propagation is by stem-cuttings or by small pieces of the tuber. These are obtained by leaving a small portion attached to the plant, when the larger part of the tuber is harvested.

In West Africa the average production of all kinds of yams is 6.5–11 tons per hectare, in East Africa 12.5–25 tons and in the Caribbean area 10–30 tons. If the tubers are subterranean and grown for an individual's own use they are often left in the ground until required, although there may then be some loss of flavour. Frequently, however, they are gathered and tied to a vertical framework which allows adequate ventilation while they are in store. The content of the yam tuber is similar to that of the potato except that in some species the poisonous alkaloid dioscorin is present. This substance is destroyed when the tubers are cooked. In West Africa the tubers are boiled and pounded to form a glutinous dough known as 'fufu'. This is flavoured with a spicy sauce, and with the addition of meat or fish, forms a favourite dish in that region.

## Greater Yam

**177**

*Dioscorea alata* L.

**Yam family** Dioscoreaceae

The genus *Dioscorea* comprises some 600 species distributed throughout the tropics and was named after the Greek physician Dioscorides. Almost all species form subterranean tubers which can weigh up to 50 kg (see photo). The tuber may be compact and undivided and may protrude above the surface of the ground. In that case its outer layer is composed of corky tissue which is more or less deeply grooved, breaking up the surface into irregular segments. In other cases several tubers are formed which are more or less united at the base and often club-shaped (Illustrated p. 184). These tubers are produced from the root-system at the base of the stem and are 20–70 cm in length and up to 20kg in weight. An annual stem, 30m or more high, arises from the top of each tuber, sometimes bearing potato-like aerial tubers in the leaf-axils. The leaves are arranged alternately on the stems, and are heart-shaped with several primary veins and a network of smaller veins. The flowers are unisexual, male and female appearing on separate plants in pendulous inflorescences. As far as is known, all *Dioscorea* plants are unisexual. Both male and female flowers are 2–4mm in diameter and usually white or greenish in colour. In spite of their small size, they are visited by insects which are attracted by their strong scent. Fruits in the form of berries are produced on the female plants.

There are three main centres of cultivation: E. Asia, tropical Africa and the Caribbean area, and a number of forms are grown whose systematic relationships have not been fully explained. (Some of the various growth forms and kinds of tubers can be seen in the illustration on the facing page.) Columbus learnt to know the plant in 1492 under the name of 'nyame'.

The greater yam, also known as the water yam, is one of the species most frequently cultivated in the tropics, and the name of the species (*alata*=winged) comes from the wings on the edges of its square stem. Where there is intensive cultivation artificial supports are necessary, but in mixed plantings, for example in Jamaica, the plants are allowed to climb up tall trees. The species has particularly deep-lying tubers, so that

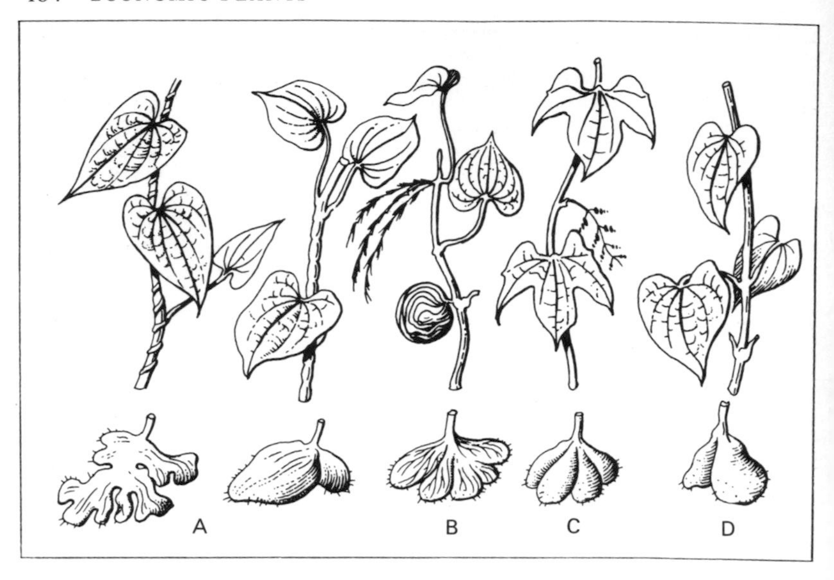


Growth-forms of A) *Dioscorea alata* (Asia); B) *D. bulbifera* (Asia); C) *D. trifida* (Central America); D) *D. cayenensis* (Africa).

they are well protected against pigs and other animals that might grub them up. The plant usually produces a single, but very large tuber, which may be divided into several parts. The yield from an individual plant is, on average, 6–10kg.

The greater yam is of asiatic origin, but during the centuries it has spread throughout all tropical regions. Portuguese sailors brought tubers to the West Indies in the 16th century, as part of their ship's provisions. *D. alata* is not so popular in Africa because of its lower starch content. The species preferred there is the yellow yam, *D. cayenensis*, which has many cultivated forms.

## Yautia **178**

*Xanthosoma sagittifolium* (L.) Schott

**Arum family** Araceae

Yautia has been cultivated for centuries in the tropics of the New World. It is also known as 'tannia' but it is better to call it 'yautia' to avoid confusion with *Colocasia esculenta* which is also called 'tannia'. The rootstock is rich in starch and, in cultivated forms, bears only a few large leaves at its top. They consist of a stalk, 2m long and channelled on the inner side, and a blade up to one metre in length. The leaf-blade is usually arrow-shaped with three main veins, one ending in the tip of the leaf and the other two running down to the pointed lobes at its base. Numerous strong lateral veins branch off, more or less at right angles, from the main veins. The leaves are glossy on the upper side and a dull bluish green beneath. Wild forms usually have many more leaves and these are smaller too. Apart from the uniformly green-leaved forms there are cultivated varieties with purple stalks and veins whose leaves can be used as a vegetable. The inflorescence is typical of the Araceae and consists of a spathe which encloses the spike of flowers known as the spadix. The female flowers are arranged at the base of the spadix in a geometrical pattern, but rarely develop into berries.

Species of the genus *Xanthosoma* are found growing wild in the West Indies, and from Mexico through Central America to Brazil. *X. jacquinii* is especially abundant as a wild plant in Mexico, Colombia, Venezuela and in the West Indies. *X. violaceum* is so called because of its purple leaf-stalk and main veins of the leaf-blade.

Yautia tubers weigh up to 2kg and are rich in starch, but have the disadvantage that they contain needle-shaped crystals of calcium oxalate (raphides), which are unpleasant to eat. Some forms are considered to be poisonous in their raw state. In the Canary Islands, where some particular varieties can be grown, the tuberous rhizomes are only sold after they have been cooked.

At the time of the slave trade, prisoners on the galleys were forced to chew parts of the raw tubers. The raphides then caused damage to the mucous membranes in their mouths and poisonous substances entered the wounds producing swelling of the tongue and mouth causing extreme pain.

Yautia is mainly grown in tropical regions of the New World and there are large plantations, especially in Cuba. It requires an abundance of light, and in sunny places produces considerably larger tubers than in the shade. Many of the native varieties are only partially cultivated. Before being eaten, the tubers are treated with earth and ashes in order to remove any corrosive and slimy constituents present. With properly controlled breeding, the quality of yautia tubers could be greatly improved.

## Taro, Coco-yam, Dasheen

**179, 180**

*Colocasia esculenta* (L.) Schott

**Arum family** Araceae

The taro plant forms an erect, tuberous rootstock with ring-like scars on the surface where leaves have fallen off. The tubers are up to 4kg in weight (Photo 180) and produce runners just above the leaf-scars with a small tuber at the end (see illustration overleaf). The small tubers are used primarily for vegetative propagation. The tubers contain 15–20% carbohydrate, 3% protein, and up to 1.7% sugar. The taro is a perennial swamp plant and requires a warm and damp climate.

The species occurs wild in Burma and Assam, and has been cultivated for 2000 years in S.E. Asia. Its principal centre of cultivation is in Polynesia, where some hundreds of cultivated varieties are grown. Another important centre is India, and from here the taro spread westwards to Africa. During the period of the slave-trade, negroes brought the plant to the Caribbean, where it is still known by its African name 'eddo'.

The taro plant is very variable. The tuberous rhizome sends up long-stalked, heart-shaped to arrow-shaped leaves which resemble elephants' ears (Photo 179). Their stalks are 7–25cm long and the blade is up to 1m in length. The leaves vary in colour from very dark green, almost black, to greenish yellow. Both stalk and blade can be spotted or striped in all kinds of ways, and there are forms where the leaf is dark on the upper side and pale beneath. Flowers are rarely produced, but when they are, the inflorescence consists of a white spathe and a spadix which has female flowers at the base, sterile ones above them, and male flowers at the top. The glossy berries are up to 5mm across with a single seed.

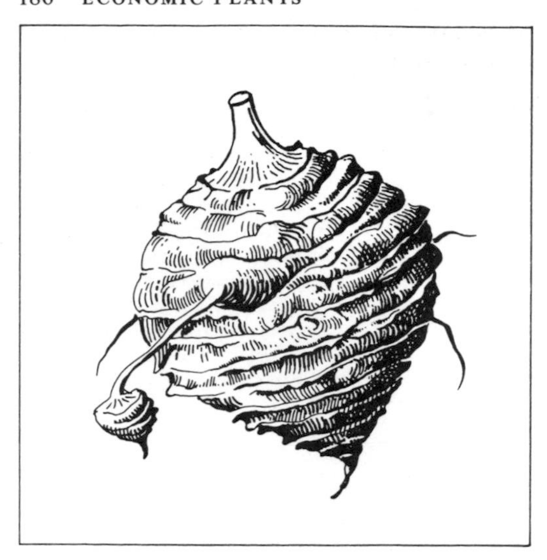

Main tuber and secondary tuber of the taro (*Colocasia esculenta*).

The tubers are rich in starch and contain not only mucilage but also calcium oxalate crystals which are injurious to the skin of the mouth. The colour of the tubers varies from pure white to dirty grey, reddish brown and bluish purple. They are also frequently used as food for pigs. The leaves are rich in protein and form an important ingredient of 'callaloo', the highly esteemed national dish of Trinidad. Yields vary between 1 ton per hectare on the Ivory Coast and 26.5 tons in Egypt, and are used in the regions of cultivation to serve the needs of the local population. In 1972, the total world yield was 3.9 million tons. If the tubers have not been properly prepared before eating, the sharp crystals of calcium oxalate can scratch the throat and also cause indigestion. Consequently the water in which they are cooked should be changed several times. A kind of flour can also be made from taro tubers.

## Breadfruit

**181, 182**

*Artocarpus communis* Forst.

**Mulberry family** Moraceae

The breadfruit is one of the plants which store starch in their fruits. It thrives only on good soils in the damp and hot equatorial climate. One species, which still occurs wild in Malaya, and may have contributed to the domesticated plant, is *A. champeden*. The leaves of this species are covered with brown hairs, and the fruits, which are smaller than those of *A. communis*, have such a penetrating odour that Europeans find them highly unpleasant. The cultivated breadfruit tree which was developed from the wild form reached Polynesia in prehistoric times, and it was there that European explorers found it a valuable source of food. In 1789, Captain Bligh had 1000 young breadfruit trees loaded on to his ship 'Bounty' at Tahiti, but on the return journey the famous mutiny took place. However, after his rescue the captain undertook another breadfruit expedition, and this time he was successful in transporting trees to the West Indies. Nowadays they are grown as ornamental plants and for food throughout the whole of the tropics.

The breadfruit reaches a high of 15–20m and has very decorative leaves up to 70cm long and 40cm broad, glossy on the upper side and usually deeply pinnately lobed. The leaves are normally evergreen, but become deciduous in monsoon regions or under semi-arid conditions. A milky juice is present in all parts of the tree. The inflorescences are unisexual, the male flowers forming a stout, stalked, catkin-like structure up to 20cm long, and the female arranged to form a globe-shaped mass with hundreds of individual flowers. When the flowers develop into fruits their bracts and also the top of the common stalk become fleshy and the whole structure forms a multiple fruit with a warty surface, 20–30cm in diameter and weighing up to 2kg (Photo 181). The fruit contains 20–22% carbohydrate and 1–2% protein. There can be as many as three harvests a year, and a single tree can produce about 50 fruits. There are both seeded and seedless forms, and the seeded forms, known as breadnuts, contain 16–24 nuts, about as big as chestnuts and similar in taste when roasted. The fruits are usually harvested when green. As they ripen they turn golden yellow and have a rather tart taste.

The breadfruit begins to fruit when it is five years old and continues to bear for 60–70 years. The seedless forms can be boiled and made into soup. When roasted they supply a flour which can be used for making bread. Unripe fruits can be dried, cut in slices and baked to make biscuits which can be stored for later use. The ripe pulp is often kept in pits where it ferments and acquires the consistency of cheese. This is considered a favourite article of food in Polynesia. The bark of the tree produces a fibre which is used for weaving and binding, and the timber is employed in boat-building.

## Jackfruit

**183, 184**

*Artocarpus heterophyllus* Lam.

**Mulberry family** Moraceae

Like the breadfruit, the jackfruit also produces large fruits which are rich in starch. Its native country is southern India, where it can be found wild in the hot coastal districts of Kerala, and is known as 'jaca' by the local inhabitants. The tree has been cultivated for centuries in India and Ceylon, and spread from there first of all to Melanesia and East Africa. Then in the 16th century the Portuguese brought it to Brazil and the Caribbean islands, where it was used long before the introduction of the breadfruit.

The jackfruit reaches a height of up to 25m, but in contrast to the breadfruit, it has entire, obovate leaves. Because of its leaf-shape it has also been known botanically as *A. integrifolius* (=with undivided leaves). It does not need such a hot climate as the related species. A further distinguishing character is the position of the inflorescences. In the breadfruit these are produced at the ends of the branches, while in the jackfruit they appear from the trunk and older branches. This condition is called 'cauliflory'.

The jackfruit too, has unisexual inflorescences, which, in the case of the female, develop into huge multiple fruits, up to 30cm in diameter, almost 1m long, and weighing as much as 50kg (Photos 183, 184). These are highly prized, both in their raw state and as a cooked vegetable by the native peoples of India and East Africa. When the fruit is eaten raw, the latex canals and fibrous strands connecting the individual carpels may cause slight digestive upsets. The fleshy fruit contains numerous seeds, up to 5cm long, which are very tasty when roasted. For many people the jackfruit is an important basic food.

The fruit contains a foul-smelling substance (caproic acid) which can be removed to a large extent by placing it over-night in salt water. The seeds resemble edible chestnuts and contain up to 39% carbohydrate and 6% protein. A single tree can bear 220–260 fruits, each weighing 20–30kg, so its food production is therefore extremely high. The tree is also highly regarded for its valuable timber, mainly used for boat-building, especially in India. In tropical regions of Africa there is a related plant known as the African breadfruit, *Treculia africana*, which also has large fruits. The seeds of this species can be ground to provide flour.

## Sugar Cane **185, 186, 187**

*Saccharum officinarum* L.

**Grass family** Gramineae

Sugar cane is a tropical food plant whose area of cultivation corresponds to a large extent with the distribution of the palms (see map on p. 13). Consequently it is possible, under favourable conditions, to cultivate the plant outside the tropical belt. The home of the sugar cane was previously thought to be India, but it has now been ascertained that its ancestors are to be found in New Guinea, where *S. robustum* still occurs as a wild plant. This species is used even today by the Papuans as a starting point for selecting suitable varieties.

Cultivation of sugar cane in the tropics goes back several thousand years. Alexander the Great learnt to know the plant on his campaigns, and Marco Polo reported on its cultivation in China in 1272. Between AD 700 and 900 the Arabs brought sugar cane to the mediterranean region, and Spaniards and Portuguese introduced the plant, via the Canary Islands, into the West Indies and also Central and South America. The Dutch laid out the first plantations in Indonesia. Independently of this, sugar cane had already spread long before from its home in New Guinea to all parts of the Old World tropics. A century ago, sugar was still a luxury commodity, sold only by chemists, but since 1910 cane sugar production has risen by leaps and bounds. It has a clear lead over the sugar beet industry and accounts for 55% of the total sugar production in the world.

Being a tropical plant, sugar cane requires an annual average temperature of about 18°C or more. Rainfall must be at least 1000–1250mm, otherwise irrigation is necessary. The ground must not, however, become waterlogged. For successful cultivation a well-drained soil with abundant nutrients and humus is required, such as the rich red earth of Cuba.

Sugar cane is a reed-like grass, 5–9m high, and can live for up to 20 years (Photo 185). It has a creeping rhizome which sends up stems sharply divided into nodes and internodes (Photo 186). The stems reach a diameter of up to 7cm, and the cells composing the soft interior are full of sugary sap. Just above each node is a band of incipient roots and a narrow growth ring. Where the node is just above the surface of the ground the roots often begin to grow out from the stem, and if the stem is at an angle the growth zone allows it to achieve an upright position. The leaves are up to 2m long and are arranged alternately in two rows on either side of the stem. They have a hard edge, due to a deposit of silicic acid. The highest sugar content is at the base of the stem and can be as much as 20%.

A total of 12 species of sugar cane have been described, of which only three can now be upheld. Nowadays, cultivated plants are almost always complex polyploids. Some of these never flower, and others flower only in the narrow tropical belt, since they are 'short-day' plants. The inflorescence is a large and easily visible panicle (Photo 187).

Vegetative propagation of sugar cane. A piece of stem (1) is placed in the ground. The shoot produced (2) in the second year is genetically identical to stem 1. Planting a piece of stem 2 would give rise to shoots 3 which are genetically identical to stem 2, and so on.

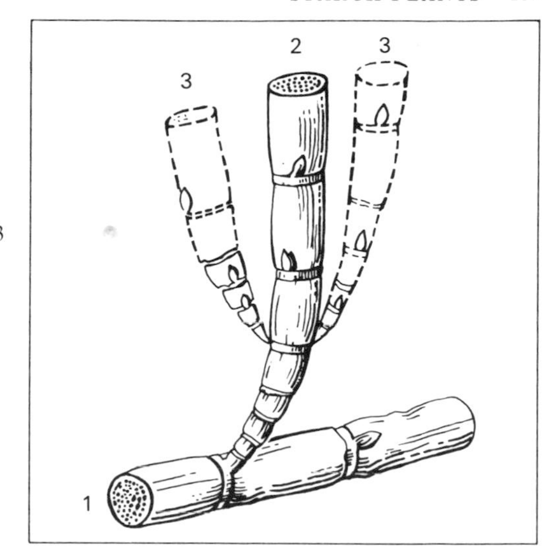


Harvesting begins when the leaves turn yellow or when core punches show that there is a sufficiently high sugar content. As a rule this occurs 10–14 months (or in extreme cases up to 24 months) after planting out. The hard stems are cut by hand, and must be brought to the factory as soon as possible after the leaves have been removed, since the sugar rapidly deteriorates in the high temperatures. If the stubble is not ploughed up, the plant can produce four to eight crops with suitable manuring. Propagation is carried out by placing stem cuttings, each consisting of two or three internodes (see illustration above) in the ground. These soon produce roots and new stems.

Yields vary between 3 tons per hectare in Yemen and 169 tons in Ethiopia, with an annual average of 53 tons. The sugar content is 7–20%. In addition there is 0.4–1.4% fructose and up to 2% lactose which renders crystallisation more difficult. A table showing cane sugar production appears on p. 190.

To extract the sugar, the canes are cut up, passed several times between rollers and crushed. The juice is first of all filtered and purified mechanically to remove any insoluble or suspended matter. It is then heated to concentrate it prior to crystallisation. The crystallised brown sugar is separated from the molasses by centrifugal means and subsequently refined. The remaining crystal-free molasses still contains 30–40% of sugar and is used as animal feed and for producing alcohol. The best rum made from this comes from Jamaica.

The way of life of the Papuans on New Guinea still resembles that of the stone age and it is centred round the sugar cane. They always take a piece as a reserve when they go hunting or fishing. They have also obtained by selection a mutant with a cauliflower-like inflorescence which they use as a vegetable. Sugar cane is also used as a roof covering for their huts.

**Production of Sugar Cane**
(in million tons)

| *Country* | *1975* | *1978* | *1980* |
|---|---|---|---|
| India | 144.3 | 181.6 | 128.8 |
| Brazil | 91.4 | 129.2 | 148.4 |
| Cuba | 55.2 | 66.4 | 68.0 |
| China | 43.0 | 47.1 | 31.7 |
| Mexico | 34.4 | 34.5 | 34.5 |
| U.S.A. | 25.7 | 24.5 | 25.6 |
| Australia | 22.0 | 21.5 | 24.0 |
| Pakistan | 21.2 | 30.1 | 28.6 |
| Colombia | 21.1 | 23.1 | 26.0 |
| Philippines | 20.8 | 20.8 | 20.9 |
| South Africa | 16.8 | 19.5 | 14.0 |
| Argentina | 15.6 | 14.6 | 17.2 |
| Thailand | 14.6 | 19.0 | 12.6 |
| Indonesia | 13.1 | 15.0 | 17.1 |
| Dominican Rep. | 9.3 | 10.8 | 9.9 |
| Peru | 9.0 | 8.4 | 5.7 |
| Egypt | 7.9 | 9.2 | 8.8 |
| Ecuador | 7.7 | 7.5 | 6.6 |
| Bangladesh | 6.7 | 6.7 | 7.0 |
| Venezuela | 5.5 | 5.1 | 5.0 |
| Guatemala | 4.9 | 5.4 | 5.6 |
| Mauritius | 4.3 | 6.2 | 4.5 |
| Guyana | 3.6 | 4.2 | 3.8 |
| Jamaica | 3.6 | 3.3 | 3.0 |
| El Salvador | 3.2 | 3.2 | — |
| Puerto Rico | 3.2 | 2.6 | — |
| Bolivia | 2.4 | 3.2 | — |
| Other countries | 50.0 | 58.6 | 70.1 |
| Total | 660.5 | 781.3 | 730.7 |

# Oil and Protein Plants

## Groundnut, Peanut, Monkey-nut **188**
*Arachis hypogaea* L.

**Pea family** Fabaceae

The groundnut is a plant which can be grown in regions with a low rainfall, although it also thrives in the damp climate of the tropics. Its cultivation has expanded considerably during the last hundred years. Even in pre-Columbian times it was being grown in the region between the rivers Paraná and Paraguay. It was first described in 1547 by the Spaniard Oviedo and reached Africa in the 16th century through the slave trade. The Spanish brought the groundnut to the Philippines, and at the beginning of the 18th century it was being cultivated in India and China, and later in Virginia, too. Wild forms of the groundnut still exist today in the subcontinent of South America, between the Atlantic coast and Bolivia. They occur only sporadically and belong to various species.

The groundnut is an annual which begins to produce branches even from the axils of the seed-leaves (cotyledons). As a rule it grows in a prostrate position, but there are also crosses between creeping and upright forms. The stems are 30–50cm long, and bear pinnate leaves, arranged alternately, which have well-formed stipules at their base. The leaves always consist of two pairs of obovate leaflets. Two to six small yellow flowers arise in the axils of the lower leaves. They are typically papilionaceous in structure, and have eight fertile and two sterile stamens. After self-pollination has taken place, the base of the ovary begins to grow. At the same time it bends round and pushes the fertile portion of the ovary into the ground (see illustration below). There it remains, in a horizontal position, for four or five months, eventually developing into a nut-like pod.

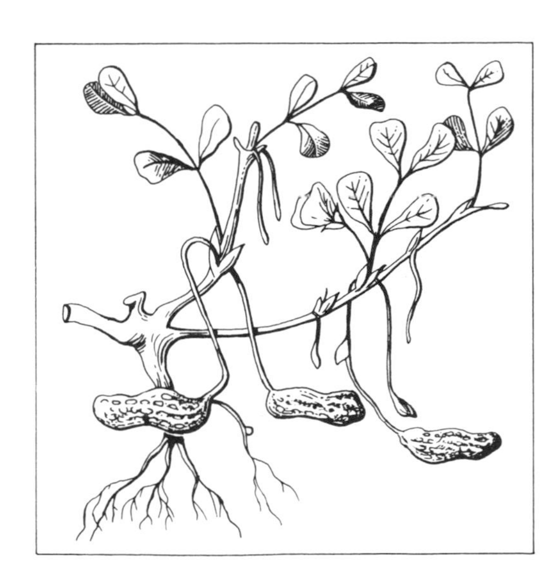

Part of a groundnut plant (*Arachis hypogaea*) with its subterranean fruits (geocarpy).

The ripening of a fruit under ground is called 'geocarpy'. The outer surface of the fruit is covered with a network of raised veins, and inside are the seeds, each enclosed in a paper-thin, red seed-coat. The seeds have two large cotyledons where nutrients are stored.

The groundnut grows best in a temperature of 25°–28°C with about 500mm rainfall. As it has a tap-root, and the fruits ripen below the ground, it prefers a light soil. This does not have to be particularly rich in nutrients, since the plant can make use of the free nitrogen in the air with the aid of bacteria in its root-nodules. At harvest-time, the tap-root is cut off and the whole plant with its pods is taken up. The plants are then dried, reducing the water-content of the seeds from 40% to 10%. The dried stems are rich in protein and are used as animal feed. A table showing yields from different countries appears below. These statistics show a comparative rise in production during recent years. A third of the yield is used by the producer-countries themselves.

In addition to their high protein content, amounting to 24–35%, groundnut seeds also contain 42–52% oil, now one of the most important of cooking oils. In order to extract the oil, the seeds are ground, warmed and treated with the solvent hexan. The oil obtained is used not only for cooking but, mixed with other oils, is employed in preserving fish and especially in the manufacture of margarine. In India the seeds are

**Production of Groundnuts**
(in thousand tons)

| *Country* | *1975* | *1978* | *1980* |
|---|---|---|---|
| India | 6755 | 6200 | 6400 |
| China | 2891 | 2883 | 3692 |
| U.S.A. | 1750 | 1809 | 1042 |
| Senegal | 1476 | 1021 | 500 |
| Sudan | 931 | 850 | 810 |
| Indonesia | 630 | 687 | 750 |
| Brazil | 441 | 325 | 483 |
| Burma | 411 | 450 | 494 |
| Argentina | 375 | 370 | 294 |
| Zaire | 308 | 339 | 323 |
| Nigeria | 280 | 700 | 570 |
| South Africa | 270 | 318 | 345 |
| Mali | 227 | 146 | 183 |
| Uganda | 182 | 208 | 200 |
| Malawi | 165 | 100 | 170 |
| Cameroun | 165 | 164 | 250 |
| Gambia | 150 | 105 | 105 |
| Thailand | 142 | 170 | 130 |
| Central Africa | 133 | 133 | 124 |
| Rhodesia | 125 | 120 | 84 |
| Mozambique | 80 | 80 | 90 |
| Niger | 42 | 74 | 100 |
| Other countries | 1499 | 1625 | 1762 |
| Total | 19428 | 18877 | 18901 |

boiled to form a mash, mixed with five or six parts of water, and strained to produce the so-called 'groundnut milk'. Peanut butter is made from a mixture of roasted and unroasted nuts. These are crushed into a uniform mass to which groundnut oil, soya flour, honey, malt and even cheese are added. The taste can vary considerably depending on the relative proportions of the ingredients.

## Soya Bean

189

*Glycine max* (L.) Merr.

**Pea family** Fabaceae

The soya bean is a plant of subtropical origin, but by selection it has developed into an agricultural food-plant of tropical latitudes also. Its native region can no longer be established for certain, although it must have come from eastern Asia north of the 40° parallel. The ancestral species is thought to be *G. ussuriensis* which is found wild in the region between the Yangtse-kiang and Korea, and also in Japan and Taiwan. The principal areas of cultivation are in eastern Asia and North America between latitudes 35°–40° north. In the tropics cultivation is continually increasing.

The plant has not been cultivated for as long as is generally thought. There is no evidence that it was grown in China earlier than 800 BC. Nevertheless it is surprising that this valuable food-plant has only been cultivated in Europe and America since the 19th century. Part of the reason for this was that the particular race of bacteria which live in its root-nodules was not at first present outside the Asiatic region.

In its general requirements the soya bean behaves very much like maize, needing high temperatures in summer and autumn to ripen properly. Its growth is aided by light, neutral soils and moderate rainfall, especially when the seeds are ripening. It is a bristly hairy annual and usually grows in a bushy manner up to 80cm high. But there are primitive forms with thin, prostrate stems reaching a length of up to 2m. The plant produces a tap-root which is strongly branched and bears numerous nodules on the fine lateral roots. In these nodules live bacteria called *Rhizobium japonicum* which have the power of fixing atmospheric nitrogen and making some available for the host plant. The pinnate leaves of the soya bean are arranged alternately, and are composed of usually three, occasionally five, relatively large, ovate, pointed leaflets. The racemes, which arise in the axils of the leaves, are made up of numerous, very small, bluish or white papilionaceous flowers. After successful self-pollination, brownish yellow, hairy pods develop which contain two or three more or less globose seeds. These may vary in colour from yellowish white to brown or black (Photo 189). Their size may also vary considerably.

The seeds are especially valuable because they contain up to 48% protein, 24% carbohydrate, and 19% oil. Yields vary between 0.8 and 3.5 tons per hectare in a year, and details of production are given in the table on p. 194. In the USA harvesting is usually carried out by combine harvesters, when the leaves have turned yellow but before the pods burst. In eastern Asia the plants are allowed to reach the same stage whereupon they are cut with a sickle, left to dry for a few days and then threshed.

Because of their high protein, starch and oil content, soya beans can be used for a variety of purposes. After the extraction of oil, which is primarily used in the manufacture of margarine, the residue, rich in protein and carbohydrates, is processed further in various ways. In eastern Asia, soya flour can be made from it and used as a basis for soups and sauces and as an ingredient in bread and pastries, also 'soya milk' which can take the place of cows' milk. With the help of bacteria and the lower fungi, the protein from the soya bean can be induced to ferment, giving rise to different kinds of vegetable curd. In the trade they appear variously as 'miso', 'tofu', 'temphe', 'natto' and 'sufu'. Other products of the soya bean are soya yoghurt and a form of condensed milk.

In recent years a form of artificial meat has been produced from soya protein. The protein is pressed through narrow tubes and collected in vats where it solidifies. The addition of a binding agent, flavouring and colouring results in a product known as TVP (textured vegetable protein) which resembles meat in consistency and taste, the flavour matching that of a variety of meats.

By adding soya flour to flour made from maize, cassava etc., the protein content of bread and other bakery products can be considerably improved. This is of great significance, especially in over-populated regions of the tropics, where people are suffering from a lack of protein. Young soya beans can be preserved or deep frozen and can be used with young maize kernels and rice to form 'succotash', a very nourishing dish. Soya bean seedlings are of particular dietary value, since after germination they contain various vitamins in high concentration. They form a part of the Indonesian dish 'rijstafel'.

In spite of various pests, cultivation of the soya bean has risen enormously in the last few decades, and in the United States, for example, areas of cultivation are continually being enlarged. At the same time, progress is continuing to be made in producing varieties suitable for the tropics.

**Production of Soya Beans**
(in thousand tons)

| *Country* | *1975* | *1978* | *1980* |
|---|---|---|---|
| U.S.A. | 42114 | 50149 | 49454 |
| China | 12662 | 13257 | 10026 |
| Brazil | 9892 | 9800 | 15153 |
| Soviet Union | 780 | 680 | 470 |
| Mexico | 699 | 324 | 299 |
| Indonesia | 590 | 530 | 600 |
| Argentina | 285 | 2500 | 3500 |
| Canada | 367 | 475 | 713 |
| South Korea | 311 | 293 | 216 |
| North Korea | 290 | 320 | 340 |
| Other countries | 1480 | 1904 | 2710 |
| Total | 69670 | 80232 | 83481 |

# Oil Palm

**190, 191**

## *Elaeis guineensis* Jacq.

**Palm family** Palmae

Among the various sources of oil for the world market, the oil palm is growing in importance, for its crops of fruit increase year by year. The generic name of the palm is derived from the Greek word 'elaion' (=oil), while the name of the species refers to its country of origin. Its home is the rain-forests of the Gulf of Guinea, where the closely related species *E. ubaghensis* can still be found in the wild. Fruits of the palm have been collected by the inhabitants of this region since ancient times. The plantations remaining at the turn of the century consisted of oil palms which had germinated naturally and were left to mature when the primeval forest was thinned out.

In contrast to the coconut palm, the oil palm has a shorter trunk, 6–15m high, and the leaf-bases remain attached to it for a long time, providing niches where epiphytes can become established. The trunk is crowned by 20–30 pinnate leaves, reaching 3–6m in length. The midrib is armed with spines, especially at the base. The compact, densely branched inflorescences arise from the axils of the leaves and are composed of unisexual flowers. After pollination by the wind, the female flowers develop into fruits as large as plums (Photo 191). Beneath the smooth outer covering is the fibrous, orange-coloured pulp which surrounds a hard kernel. Inside is the seed, rich in oil like the pulp. The oil palm begins to produce flowers and fruit from its fifth year, when it is still in the rosette stage, and it continues to do so until it is 50 or even 80 years old.

The bunches of fruit weigh up to 20kg and are harvested by hand. On the day of harvest they must be heated by steam in order to prevent the oil present in the outer layer of the fruit from splitting into glycerine and fatty acids.

The oil is first obtained by crushing the fruits. 50–70% of the oil present in the fleshy outer tissue can be extracted in this way. This is called 'palm oil' and it has a variable amount of orange colouring according to the carotene content. The hard shell surrounding the seed is 4–8mm thick, and this must first be cracked before the oil can be squeezed out of the seed. This oil, known as 'palm kernel oil', contains over 50% lauric acid which is capable of producing a lather and is therefore used mainly in the soap industry. Palm oil, on the other hand, is used primarily in the manufacture of margarine but also in the making of candles. After processing it becomes a solid, yellowish fat. Yields of oil amount on average to 2.5–4 tons per hectare.

The first oil palms were introduced into Java by the Dutch in 1848, and in 1850 palm oil first reached Europe. From its home in central Africa, the oil palm has spread to all parts of the tropics as a cultivated plant, and today the places with the highest production are in Malaysia. At the present time, the oil palm supplies more than 12% of world demand for oil. Other parts of the palm are useful too, for example, the leaves, which are a source of fibre. Palm wine is made by cutting the male inflorescences, and collecting the sugary sap which oozes out from them. This is then fermented and distilled to produce arrack.

## Akee 192, 193

*Blighia sapida* Koenig

**Soapberry family** Sapindaceae

The akee tree was given its Latin generic name in honour of Captain Bligh, who introduced the breadfruit to the West Indies (see p. 186), but it was slave traders who, at the end of the 18th century, brought it to the West Indian islands from its native country of West Africa. The akee tree is very popular in Jamaica and can be found in almost every garden.

The tree reaches a height of 15m and has pinnate leaves made up of obovate to broadly linear, shortly pointed leaflets. From the fifth year onwards it produces racemes of white, five-petalled flowers, which are used in Africa for making perfume because of their pleasant scent. The superior ovary, consisting of three united carpels, develops into a somewhat triangular capsule, as large as one's fist and tinged red on the outside (Photo 193). Each of the three cells splits at maturity to expose an ovoid, shiny black seed, the lower part of which is enclosed in a fleshy, cup-shaped, cream-coloured aril. The oily aril has a nutty taste, and is usually cooked although it can be eaten raw. It is cooked with salted fish to provide the national dish of Jamaica. The pink tissue joining the aril to the seed must be carefully removed since it contains hypoglycin and is highly poisonous.

Unripe or overripe fruits are also dangerous. The tree is evergreen and is useful in providing shade. At Christmas time the fruits hang like red lanterns from the branches.

## Brazil Nut, Pará Nut 194

*Bertholletia excelsa* Humb. & Bonpl.

**Brazil Nut family** Lecythidaceae

The Brazil nut is named after its country of origin, and its other name refers to the town on the estuary of the Amazon which is now known as Belém. Botanically it is not a nut, but the thick-shelled seed of a capsule produced by one of the largest trees in the primeval forests of South America.

The Brazil nut is a magnificent tree, 30–40m high, and is native in the tropical rain-forests along the Amazon, Rio Negro, Rio Vaupés and Orinoco. It has never been cultivated, all the Brazil nuts of commerce being collected in this region. The tree only grows on the higher, firmer ground, *i.e.* in those places not subject to river-flooding, and it requires an average annual temperature of 28°C. Its natural area of distribution takes in Bolivia, Peru, Ecuador, Brazil, Colombia and Venezuela, but the nuts are exported only from Brazil. In order to gather stocks for export, journeys must be made to the forest localities where fallen, unopened fruits can be found. It is therefore remarkable that there is a total export of 50,000 tons per year, particularly in view of the fact that all the fruits have to be opened to obtain the nuts. In its natural habitat, of course, the fruits are opened without man's help, since the agouti, a large rodent, gnaws through the woody capsule-wall.

The Brazil nut tree bears oblong-ovate leaves, up to about 60cm long, which form a dense canopy. At flowering-time it produces panicles of large, yellow flowers which, like the cannonball tree, have numerous stamens. The flowers have an inferior ovary which develops into a large, brown, semi-globular capsule, with an aperture at the end closed by a woody plug. The capsules are usually 10–15cm in diameter but can be as much as 30cm and then weigh about 3kg. Inside are 15–30 triangular, curved nuts with

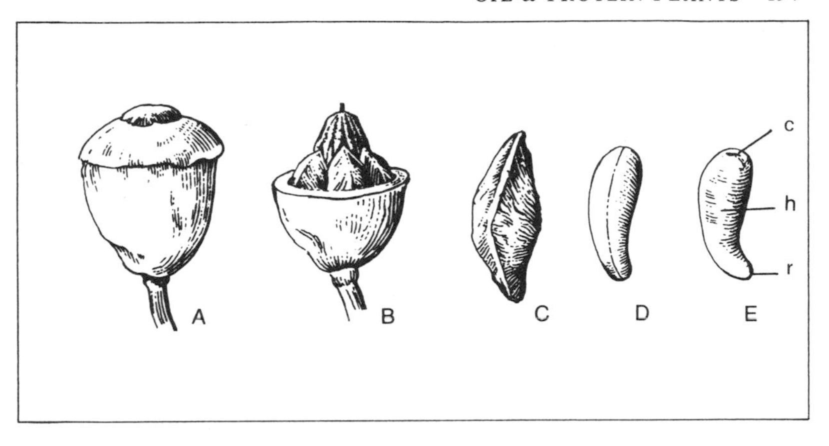


Brazil nut (*Bertholletia excelsa*): A. Capsule with plug; B. Capsule halved, showing seeds; C. Seed; D. & E. Embryo (c=cotyledons, h=hypocotyl, r=radicle).

a hard, woody, wrinkled shell (see illustration above). The shell encloses the embryo, which, like the cotyledons, is very small. On the other hand, the section between the cotyledons and the radicle, the hypocotyl, is enormously enlarged. This is really the main storage tissue and contains about 60% oil and 18% protein. The tree begins fruiting after it is ten years old and produces 100–500 capsules at a time.

Closely related to the Brazil nut tree is the genus *Lecythis*, which contains a number of species with similar, though smaller fruits. Their capsules resemble a pot with a lid, and, in fact, are called 'monkey pots'. These also contain edible seeds which are known as 'paradise nuts'. The best known species of Monkey Pot tree is *L. usitata*, though there are many others which might be considered as a basis for future commercial varieties.

## Coconut **195, 196, 197, 198**

*Cocos nucifera* L.

**Palm family** Palmae

The distribution of the coconut palm corresponds broadly with the region between the Tropic of Cancer and the Tropic of Capricorn, so it can be regarded as a typical tropical economic and also ornamental plant. There is a saying that there are as many uses for the coconut as there are days in a year. Its fruits have been used for 4000 years as an indispensable source of nourishment by the inhabitants of the South Seas and southern Asia. Since 1740 it has been systematically cultivated by the Dutch and the Portuguese.

The home of the coconut palm is thought to be Polynesia, though some authorities argue that it originates from South America. What is certain is that the coconut palm is almost entirely restricted to coastal regions, and it is often the first glimpse of the tropics for the traveller as his ship approaches land (Photo 198). Of course, in exceptional cases it can penetrate inland along river-banks to a distance of over 150km, and in Africa and Peru it is found at higher altitudes. However, it is really a salt-tolerant, tropical coastal plant, whose buoyant fruits have been proved capable of drifting up to 4500km with the help of ocean currents and still remaining viable.

This very attractive palm grows unbranched to a height of up to 30m and, on beaches, usually stands on a kind of pedestal consisting of the swollen base of the trunk and adventitious roots. The trunk is crowned by a tuft of 20–30 pinnate leaves which may reach a length of 6m and a weight of 15kg. The paniculate inflorescences arise from the axils of the leaves and consist of a few female flowers at the base and, above them, 200–300 male flowers. All the flowers have three sepals and three petals. The female flowers have an ovary composed of three united carpels which develops, in the course of 12 months, into a fruit (Photo 196), called botanically a 'drupe'. It is only the inner part of this fruit, consisting of the kernel enclosed in its hard shell, which is exported to countries outside the tropics. The fibrous layer, lying between the shell and the smooth hard outer skin of the fruit (Photo 197), is sold as 'coir' for making ropes and coconut matting.

The seed is situated inside the kernel, and consists of a small embryo and a white, fleshy layer, the endosperm, which, when ripe, forms the inner wall of the kernel and is known as the 'meat' of the coconut. The hollow centre is partially filled by 'coconut milk', a clear liquid which oozes from the endosperm as it changes from a liquid to a solid substance. The embryo is only a few millimetres long and is embedded in the solid tissue at the end of the fruit under one of the three 'eyes'. This remains soft and easily penetrable, while the other two eyes become hard and woody. During its germination and for several months afterwards, the young coconut plant absorbs the nutrients

**Production of Copra**
(in thousand tons)

| *Country* | *1975* | *1978* | *1980* |
|---|---|---|---|
| Philippines | 2020 | 2600 | 2000 |
| Indonesia | 885 | 950 | 991 |
| India | 314 | 329 | 378 |
| Sri Lanka | 203 | 146 | 126 |
| Malaysia | 183 | 207 | 231 |
| Mexico | 145 | 160 | 120 |
| Papua-New Guinea | 135 | 132 | 140 |
| Mozambique | 63 | 75 | 68 |
| Thailand | 41 | 48 | 51 |
| New Hebrides | 35 | 38 | 50 |
| Tanzania | 27 | 27 | — |
| Solomon Is. | 25 | 28 | — |
| Fiji | 24 | 30 | — |
| Vietnam | 23 | 24 | 39 |
| French Polynesia | 22 | 24 | 37 |
| Other countries | 221 | 214 | 321 |
| Total | 4368 | 5032 | 4552 |

present in the surrounding tissue and becomes a spongy mass almost filling the interior of the kernel (Photo 195: germinating coconut).

As it is a typical tropical plant, the coconut palm requires an average annual temperature of 27°C and a rainfall of 1200–2000mm. It is not only tolerant of salt but likes an abundance of light. It can live to be over 100 years old and produces its best crops of fruit, 50–80 per tree, from its twelfth to its fortieth year. Harvesting is carried out by men or trained monkeys climbing the trees, or by using knives attached to long poles. When the outer layers have been removed, the fleshy, white inner tissue, the 'meat' of the coconut, is dried and chopped to form 'copra'. This contains 63–70% of oil, and is used as a cooking oil and also in the manufacture of margarine. Details of copra production are given in the table on p. 198. Before drying, the inner tissue contains 35% fat and 9–10% sugar and it is grated and macerated for domestic consumption in the countries of origin. The leaves of the palm are used for thatching, and their midribs alone bound together as brooms. The trunk provides timber for building and for furniture. Palm wine and arrack are made from the juice tapped before and during the flowering period. From the eighth month onwards the ripening fruits provide a sweet-tasting drink which is both refreshing and nourishing. The hard shells supply material for heating and for the production of charcoal, and are also used to make household utensils.

# Beverages, Spices, Flavouring Plants, Food Dyes, Masticatories.

## Arabian Coffee
*Coffea arabica* L.

## Congo Coffee, Robusta Coffee
*Coffea canephora* Pierre ex Froehner
Syn: *C. robusta*

## Liberian Coffee
**199, 200, 201**

*Coffea liberica* Bull ex Hiern

**Madder family** Rubiaceae

The word 'coffee' is derived from the Turkish and Arabic languages and probably goes back to the word 'Kaffa', the name of a region in Ethiopia, for it is there that the wild form of Arabian coffee still occurs. However, the coffee industry has never been developed in Ethiopia. It appears that the plant was first cultivated between AD 1000 and 1300 in the Yemen, where it was systematically grown on irrigated hill-terraces. The Arabs took control of its cultivation and were responsible for spreading coffee throughout the world. Their enthusiasm was such that in 1511 a law had to be passed in order to prevent excessive visiting of coffee-houses. In 1550 a public coffee-house was opened in Constantinople but was soon closed by order of the sultan. The first news of coffee reached western Europe in 1576 through the German doctor and botanist

Rauwolf, who undertook an adventurous journey through the Orient, and later wrote down his experiences in a travel book. The first coffee-house in Europe was opened in Vienna in 1683, after the Turks had withdrawn from an unsuccessful siege of the town. About 1710 the Dutch started growing coffee in Java, and in the period 1750–1850 the first large plantations were established in Brazil.

The coffee plant is a tree, 2–3m high, with glossy evergreen leaves arranged in pairs on the stems. The leaves are elliptic to lanceolate in shape and pointed at the end (Photo 200). Compact groups of four to six, or, in the case of robusta coffee, up to 60 flowers are produced in the axils of the leaves. In regions with a seasonal climate they open at the onset of the rainy season turning the coffee plantations into a sea of brilliant white foam. The ovary of the flower develops into a red fruit or 'cherry', botanically called a 'drupe' since within the pulp is a hard, horny layer surrounding two seeds – the 'coffee-beans'. Each of these is wrapped in a transparent seed-coat known as 'silver skin', and

**Production of Coffee**
(in thousand tons)

| *Country* | *1975* | *1978* | *1980* |
|---|---|---|---|
| Brazil | 1263 | 1200 | 1067 |
| Colombia | 540 | 669 | 763 |
| Ivory Coast | 270 | 198 | 245 |
| Mexico | 228 | 270 | 222 |
| Uganda | 213 | 156 | 123 |
| Ethiopia | 179 | 191 | 193 |
| El Salvador | 165 | 132 | 161 |
| Indonesia | 160 | 191 | 240 |
| Guatemala | 129 | 139 | 156 |
| India | 93 | 119 | 150 |
| Madagascar | 91 | 87 | 80 |
| Zaire | 83 | 95 | 90 |
| Costa Rica | 82 | 95 | 113 |
| Cameroun | 80 | 90 | 102 |
| Ecuador | 76 | 89 | 89 |
| Angola | 68 | 54 | 40 |
| Kenya | 66 | 80 | 91 |
| Venezuela | 65 | 72 | 66 |
| Philippines | 62 | 82 | 125 |
| Dominican Rep. | 62 | 45 | 57 |
| Peru | 59 | 66 | 100 |
| Tanzania | 52 | 42 | 52 |
| Honduras | 51 | 59 | 76 |
| Nicaragua | 49 | 60 | 55 |
| Papua-New Guinea | 39 | 47 | 50 |
| Other countries | 237 | 255 | 365 |
| Total | 4462 | 4583 | 4821 |

contains up to 1.3% caffeine as well as protein and sugar. If only one ovule is fertilised, a single roundish or cylindrical seed develops, and the fruit is then known as 'peaberry'. Coffee bushes flower when three years old and from then on continue to yield well until their thirtieth year.

They thrive in regions up to 28° north or south of the equator and need an average temperature of 18°–22°C, these details being obtained from a coffee plantation in the Central American republic of El Salvador. Arabian coffee plants require a rainfall of 500–1500mm, the other two about 2000mm in a year. Arabian coffee grows best at altitudes of 600–1200m but can be cultivated up to 1700m. Liberian coffee is best suited for lowland regions, and robusta coffee, which is native in West Africa, prefers an altitude of 300–600m. Those coffee plantations situated at moderate altitudes, together with their accompanying shade-trees, give the impression from a distance of closed forest (Photo 199). There is often a conspicuous growth of epiphytes on the shade-trees, consisting of bromeliads, orchids, members of the Araceae and ferns.

Coffee 'cherries' ripen in the course of 8–12 months. As the flowering time extends over several months according to altitude, harvesting also stretches over a long period. Directly after gathering the fruit, processing begins, either by the dry or wet method. In the first case, the 'cherries' are dried in the sun (Photo 201), and then the pulp, the horny layer and the 'silver skin' are removed by machine. If the wet method is used, the fruits are washed, lightly pressed and again washed thoroughly to remove much of the pulp. Those fruits with flesh still adhering to them are placed in water-filled fermenting tanks for one to three days. During this time bacteria cause the remainder of the flesh to decompose and come away from the horny layer. The seeds, still in their protective covering, are then sun-dried, and often exported in this form. This method is considered as producing the best quality raw beans. Before being used for preparing the drink, they must of course be roasted. Yields vary between 500 and 4000kg per hectare, and details of coffee production are shown in the table on p. 200. With the rise in living standards, coffee production has also risen. A particular variety of coffee known as Mocha gets its name from Al Mukha, a port in South Yemen.

Of the three species of *Coffea* the relatively small Arabian coffee plant is the most important, producing 74% of the total world crop. 24% comes from Congo coffee trees, and the rest from Liberian coffee. Congo or robusta coffee trees are the most vigorous, and it is these that are grown especially in Brazil. Arabian coffee trees thrive best in shade, as they do in their natural habitat. Members of the pea family are mainly used as shade-trees, such as species of *Inga* and *Gliricidia*, since they are able, with the help of bacteria in the nodules of their roots, to make use of the free nitrogen in the air.

## Guaraná
*Paullinia cupana* H.B.K.

**Soapberry family** Sapindaceae

The genus *Paullinia* was named after Christian Franz Paullini, physician to a German emperor, and its fruits have the highest caffeine content of all known plants. It is a liane which grows in the primeval forests of Brazil and Venezuela, and climbs up into the crown of its supporting trees with the aid of its lateral branches. The large leaves are made up of several leaflets, 20 × 10cm, with a toothed margin. From the axils of the leaves arise paniculate inflorescences with rather insignificant flowers (see illustration overleaf). The ovary, composed of three united carpels, develops into a red capsule, the size of a hazel-nut, containing a single seed. Its embryo contains, in addition to 10.7% fat, 2.7% protein, 9.4% starch, and 49% raw fibre, more than 4% caffeine. The seeds are

Leaf and inflorescence of Guaraná (*Paullinia cupana*).

prepared for use by being ground up, mixed with cassava flour and water, and made into a paste known as 'pasta guaraná'. In its native area the paste is stirred into water to produce a refreshing breakfast drink. It has widespread appeal in Brazil, where it is bottled like lemonade and sold as 'guaraná'. The paste was formerly available in chemists' shops as a tonic. In the same way that *P. cupana* is used in Brazil, so *P. yoco* is used to make a drink by the inhabitants of the river-basins of Colombia and Ecuador.

## Tea

**202, 203**

*Camellia sinensis* (L.) O. Kuntze

**Tea family** Theaceae

The tea plant is very closely related to the ornamental shrub *Camellia* and is a member of the same genus. Its home is on mountain slopes in the Chinese province of Yunnan and in evergreen forest on the foot-hills of the Himalaya in Burma and Assam. There the plant occurs wild in the forest undergrowth up to an altitude of 2000m. Cultivation of tea began when Chinese hill-farmers started to grow tea bushes in the gardens of their houses, and both cultivation and enjoyment of tea are recorded in Chinese literature of 2700 BC. From AD 600–800 tea developed into the Chinese national drink, but in Japan it was cultivated only from about AD 1100. Tea reached Europe through the Arabs about 1550, and arrived in England about 1640. It is highly regarded in southern Asia and extensively planted in parts of India and Ceylon, as well as in southern Russia, East Africa, Argentina and Turkey.

The tea plant is a small tree, which reaches a height of 3–4m in the 'Bohea' group of cultivated forms developed in China, and up to 12m in the 'Assam' group. The leaves of the latter are broader, larger, and often paler in colour than those of the 'Bohea' group which have lanceolate, pointed, and slightly toothed leaves. When young, the leaves are covered with fine hairs. The flowers (Photo 203) resemble single forms of the Camellia. They have numerous stamens and produce a capsule which splits into three parts, each part containing a large, oily seed. Propagation is usually carried out vegetatively, by means of cuttings.

Tea plants require a warm climate of 18°–28°C, mild in winter, and with an average annual rainfall of 2000mm, as is the case in the hilly districts of Ceylon, southern India, Darjeeling, and the Azores. The bushes are pruned to maintain a convenient height, and harvesting of the tips of the shoots, together with the two or three leaves immediately below, takes place from the fourth to the twelfth year of growth. Shoots are picked five or six times annually in China and Japan at intervals of six weeks, but in India and Indonesia it can be up to 15 times or more in the course of a year. It takes 20kg of shoots, picked in 12–16 hours, to produce 4–4.5kg of dried tea. The pickings are first placed in pans in drying-houses and warm air is passed over them. This reduces the water content from 75% to 30–40%. The leaves, now soft and pliable, are passed on to rollers, prior to fermentation. This process takes place in special rooms at a temperature of 35–40°C and lasts four hours, during which time the tea acquires its particular aroma and takes on a brown colour. It is subsequently dried by means of hot air at 80°–110°C. In the production of 'green' tea, the fermentation process is omitted, the leaves being merely rolled and subjected to steam to prevent any fermentation taking place.

Then follows grading and sorting. 'Flowering Orange Pekoe' and 'Orange Pekoe' are made from only the finest leaves. Other kinds, such as 'Pekoe', 'Pekoe Souchon' and 'Souchon', are made up of coarser leaves, leaf-stalks, and pieces of stem. These contain 2.7–3.3% caffeine, 0.17% theobromine, and tannins. About a half of the total world production is used in China, India, Indonesia and Japan, all countries where tea is cultivated.

**Production of Tea**
(in thousand tons)

| *Country* | *1975* | *1978* | *1980* |
|---|---|---|---|
| India | 487 | 565 | 590 |
| China | 316 | 356 | 330 |
| Sri Lanka | 214 | 215 | 191 |
| Japan | 105 | 105 | 99 |
| Soviet Union | 86 | 111 | 125 |
| Indonesia | 70 | 73 | 92 |
| Kenya | 57 | 93 | 90 |
| Turkey | 56 | 84 | 120 |
| Argentina | 39 | 23 | 36 |
| Bangladesh | 29 | 35 | 39 |
| Malawi | 26 | 32 | 30 |
| Iran | 21 | 27 | 29 |
| Mozambique | 18 | 14 | 18 |
| Tanzania | 14 | 20 | 17 |
| Other countries | 67 | 80 | 80 |
| Total | 1605 | 1833 | 1496 |

## Cocoa

**204*, 205***

*Theobroma cacao* L.

**Cocoa family** Sterculiaceae

The generic name *Theobroma* means 'food of the gods', and the name of the species is derived from 'cachoatl', the Indian name for the plant in its native region, northern South America. Here, various species of *Theobroma* colonise the flooded parts of the dense, primeval forests at the sources of the Orinoco and the Amazon. It is not known for certain if the plant also occurs wild in the permanently damp lowland forests of Central America.

Cocoa seeds were held in high esteem by the Incas, Mayas and Aztecs. At the beginning of the 16th century they were the basis of the monetary system in Mexico and for a long time served as coinage within the distribution area of the plant. The first cocoa fruits to reach Europe were brought to Spain by Columbus, and the first proper botanical description of the plant was given by A. von Humboldt in 1806. But cocoa was already known in Europe as a drink as early as the 17th century. It was about this time that the first chocolate was made by nuns in Guanaco with the aid of sugar imported from Cuba. Because of the rising consumption of cocoa, large plantations were laid out, not only in Central and South America, but also in Ceylon, Indonesia, New Guinea and the Philippines. In Africa, the cultivation of cocoa only developed in the 19th century, but then it expanded so rapidly that today more than half of all the cocoa beans are produced there.

The cocoa plant is a small tree, which can reach a height of 15m, but in cultivation it is always kept lower than this.

The tree requires a shady environment, and a continuously damp, tropical climate, with an average annual temperature of 24°–28°C and a rainfall of 1000–6000mm. It can tolerate temperatures above 40°C, but below 15°C there is some damage to growth. The ground must be deep, damp and rich in nutrients. Cultivation is only possible within a narrow belt up to 15°, or at most 18°, north and south of the equator. The cocoa tree has evergreen, lanceolate, pointed leaves up to 30cm long, which are tinged red when young, like those of many tropical trees, due to the formation of anthocyanin. The very small, yellowish white flowers, at most only 1cm across, appear on the trunk and at the base of the older branches. This is an example of cauliflory. The ovary ends in a five-branched stigma, and is composed of five cells, each containing two rows of ovules, about 50 in all. After fertilisation, these develop into the well-known cocoa beans. They are arranged in rows inside the fruit which is 15–20cm long and resembles a cucumber. The fleshy part surrounding the seeds becomes reddish when ripe, and the outside of the fruit is greenish, then yellowish, and finally reddish brown as it ripens (Photo 204). The thin shell enclosing the seed changes from white to reddish brown as it dries.

The fruits take five to eight months to ripen, and are harvested by cutting through their hard stalks with hooked knives. To prevent the seeds germinating, the fruits must be opened. The seeds are removed, piled in heaps, and allowed to ferment. During the fermentation process, any pulp remaining on the seeds decays away, the bitter substances disappear and a brown colouring arises as a result of oxidisation. After six days, when the aroma has fully developed, fermentation is stopped. The beans are then washed and dried.

The outer shells of the seeds are removed by crushing and are used as fertilisers,

* These pictures have been misnumbered as 203, 204 in the plates.

fodder or fuel. The pulp is used to make jelly or is fermented to produce alcohol and vinegar. The seeds, on the other hand, are not roasted until they reach the consumer country. The interior of the seed, consisting almost entirely of the folded cotyledons, contains 53% fat, and is the source of cocoa butter. In the manufacture of cocoa powder, half of the fat, which also contains 14% protein and 7% carbohydrate, is removed.

The stimulating effect of cocoa products is due to theobromine, an alkaloid related to caffeine, which forms 1.2% of the dried beans and 2.3% of cocoa powder. Cocoa is especially valuable since, in addition to being a stimulant, it has a high nutritive value. Annual yields of dried beans vary between 200–500kg per hectare.

**Production of Cocoa**
(in thousand tons)

| *Country* | *1975* | *1978* | *1980* |
|---|---|---|---|
| Ghana | 396 | 255 | 255 |
| Brazil | 266 | 266 | 294 |
| Ivory Coast | 227 | 275 | 325 |
| Nigeria | 216 | 160 | 175 |
| Cameroun | 96 | 100 | 110 |
| Ecuador | 75 | 73 | 95 |
| Mexico | 34 | 34 | 35 |
| Dominican Rep. | 33 | 34 | 32 |
| Papua-New Guinea | 33 | 32 | 30 |
| Colombia | 21 | 31 | 41 |
| Venezuela | 19 | 18 | 17 |
| Togo | 17 | 16 | 18 |
| Malaysia | 15 | 22 | 34 |
| Other countries | 87 | 87 | 96 |
| Total | 1535 | 1403 | 1227 |

## Cola **206**

*Cola acuminata* (P. Beauv.) Schott & Endl.

**Cocoa family** Sterculiaceae

The cola tree is found wild as part of the undergrowth in the tropical rain-forests of West and Central Africa, and the inhabitants of this region have long used it as a masticatory. Because of the stimulating properties of its seeds it is becoming more and more important. For this reason the cola tree is not only cultivated in Central Africa but also in the West Indies, South America and eastern Asia.

The tree grows to a height of 6–15m, and has evergreen leaves, 10–15cm long and 3–5cm broad. The older leaves are oblong-ovate in shape and undivided, but the younger leaves often have one or two deep incisions at the base. The yellowish white or

Fruit of the Abata Cola tree (*Cola acuminata*).

purplish starry flowers are arranged in compact panicles at the ends of the branches or older stems (ramiflory). They are either bisexual or simply male. As the carpels are united at the base, they develop into a star-shaped compound fruit made of individual follicles which split down one side only. Inside are five to nine seeds as large as plums which are surrounded by a fleshy seed-coat and weigh 10–25g or at times as much as 50g. The seeds are known commercially as nuts, although botanically this is incorrect, and instead of having two cotyledons usually have from three to seven. These are white, pink or red at first, but as they dry they change to a reddish brown colour.

In addition to carbohydrate, sugar, fat and protein, the seeds contain an average of 2% caffeine and 0.05% theobromine. They are used to make refreshing drinks which are known throughout the world under the name 'cola'. *C. nitida* (Gbanja cola) is even more important than *C. acuminata* (Abata cola) as a source of seeds. Other species are *C. anomala* (Bamenda cola) and *C. verticillata* (Owé cola). The trees begin to flower when they are five to seven years old and bear fruit up to the age of 100 years. Each mature tree will yield a crop of 10–16kg of seeds and the total world production amounts to about 20,000 tons per year.

For centuries many people in Africa and South America have chewed cola nuts to alleviate hunger, thirst and fatigue. At first they have a bitter taste, but chewing changes the starch into sugar. At the same time, the caffeine is released and produces its stimulating effect, so the nuts, which are chewed for up to an hour, have a stimulant as well as a food value. In a number of countries, including the Sudan, the dried seeds are ground up and mixed with milk and honey to form a drink.

## Betel Palm, Areca Palm

**207**

*Areca catechu* L.

**Palm family** Palmae

The betel palm is native in S.E. Asia and is cultivated there for its seeds. The tree has a smooth grey trunk, up to 30m high, crowned by finely pinnate leaves, 3–4m long. From its sixth year it produces, in the axils of the leaves, paniculate inflorescences, which have the larger, female flowers at the base and numerous, smaller, male flowers above them. The three-celled ovary of the female flowers develops into a yellow or orange fruit as large as a hen's egg. Beneath the smooth outer skin is a fibrous layer which surrounds a single large seed, popularly but incorrectly known as a nut. Besides the nutrient tissue this contains a number of alkaloids, including 0.3–0.6% arecoline, also about 15% of tannins and red colouring matter.

The seed is first cut into slices, then lime, cinnamon and other spices added, and the whole wrapped in a leaf of betel pepper (see below). This forms the betel quid which is chewed as a stimulant by the peoples of south-eastern Asia. The tannins promote the flow of saliva, and the alkaloids have a stimulating effect. The colouring matter present turns the saliva bright red. The stimulating substances in the seed accelerate the heart and promote digestion. In addition, perspiration is increased and the gums strengthened. It is also a proven vermifuge. One side-effect is that, in the course of time, first the edges and gradually the entire surface of the teeth look as though they have been painted with black lacquer. It was Marco Polo who first brought the news of betel-chewing to Europe in 1298. This custom is still practised by more than 200 million people in India, Pakistan, Ceylon and S.E. Asia, but appears now to be on the decline.

## Betel Pepper

**208**

*Piper betle* L.

**Pepper family** Piperaceae

Betel pepper is grown for its fresh aromatic leaves, which contain 0.6–1.2% of ethereal oils including the invigorating substance eugenol. Like the black pepper, *P. betle* is a climbing plant with thickish, heart-shaped, long-pointed leaves, arranged alternately up the stems. It occurs wild in the Indian rain-forests, and needs more moisture and shade, and a deeper soil, than the shallow-rooting black pepper. The inflorescences are composed of flowers without any perianth. Only the leaves are used. These contain no alkaloids, and are wrapped round slices of the seed of the areca palm to make a betel quid (see above). The habit of betel-chewing is prevalent from Polynesia, across the Malay States to India. From there it spread, centuries ago, to Zanzibar, Madagascar and East Africa.

## Turpentine Tree

**209, 210**

*Bursera simaruba* (L.) Sarg.

**Bursera family** Burseraceae

The genus *Bursera* comprises 100 species distributed throughout the tropics of the New World. These include a number of turpentine trees which are native in the deciduous forests occurring in those regions with a seasonal climate. The best known is *B. simaruba* which is widespread in Central America and the islands of the West Indies, and is the source of gomart resin. The tree grows to a height of up to 15m and is found everywhere as a hedge-plant and street-tree. It has a strikingly smooth and shiny brown trunk, and its bark comes away in large, thin, almost transparent strips, rather like pieces of paper (Photo 210). If the trunk is damaged, a gummy resin oozes out.

The tree has pinnate leaves, composed of five or seven oblong-ovate leaflets with long-pointed tips, which are shed during the dry season. The three-petalled flowers are greenish to yellowish in colour, appear in April and May, and develop into capsules which split into three parts.

All parts of the tree are aromatic and its resin is used in lacquers and varnishes and as incense in church ceremonies. In addition, roots, bark, leaves, flowers, and young fruits are used in decoctions and teas, particularly for stomach troubles and diarrhoea, but also just as a refreshing drink.

The turpentine tree is closely related to a number of other species in Central and South America, and in the West Indies, and these have similar uses. These trees must not be confused with the Tolu balsam *Myroxylon pereirae*, which is native in Central America and southern Mexico. This is a tall, forest tree with pinnate leaves and belongs to the Fabaceae. It produces curious, one-seeded pods, broadly winged on one side and narrowly winged on the other, containing a single kidney-shaped seed. There is a related species in Peru and southern Brazil.

## Black Pepper — 211, 212

*Piper nigrum* L.

**Pepper family** Piperaceae

Black pepper has been one of the most important spices since ancient times and originates in the southern foothills of the Himalaya and the hilly districts of Assam and Burma. Because of its origin, it requires a hot, rainy climate with more than 2000mm of rain a year. It is a tropical climber, which can grow 10–15m high, attaching itself by means of adventitious roots to the supporting trees (Photo 211). Its woody stems reach a diameter of 3–5cm at the base, and the long side-shoots bear alternate, narrowly heart-shaped leaves, 8–18cm long and 5–10cm broad, on 5cm long stalks. The inconspicuous greenish flowers are in spikes 15cm in length. The flowers are composed of a large ovary and two stamens. The perianth is absent. As the flowers at the base of the spike open so early, it is impossible for self-pollination within the spike to take place.

After fertilisation, small round fruits develop, up to 5mm in diameter, which are at first green (Photo 212) and later turn red. Although these look like berries, they are regarded botanically as drupes. If the fruits are gathered before they are fully ripe, when they are just turning red, and are dried in the sun or by artificial heat, black peppercorns with a thin, wrinkled skin, are produced. White pepper, on the other hand, is obtained from fully ripe fruits. They are either placed in water to remove the thin, fleshy covering, or kept moist, covered with cloths and allowed to ferment. After three days the skin can be removed, leaving the greyish white, mild-tasting 'white peppercorns'. The substances which give the peppercorns their spicy taste are the 1–2.5% of ethereal oils with terpenes and the alkaloids piperine (5–9%) and chavicine (0.8%). Nowadays pepper is cultivated throughout the tropical regions of the world, and is usually grown up tall frames. The most important producers are India, Indonesia, Ceylon, S.E. Asia, Philippines, West Indies, Brazil, Nigeria, East Africa and Madagascar. As a large quantity is used by the producer countries, world production is difficult to ascertain, but amounts to some 100,000 tons annually. Besides black pepper, other species are grown as spices, including *P. longum* in India and *P. officinarum* in Indonesia.

## Paprika, Sweet Pepper

213, 214

### *Capsicum* spp.

**Nightshade family** Solanaceae

The genus *Capsicum* comprises 35 species with a natural distribution from southern South America, across Central America and the West Indies to Florida. The origin of the generic name is unknown, and the plant is known by different names in the various regions. A distinction should be made between those which produce spice and those which can be eaten as a vegetable. Plants were being grown as spice in the New World before the arrival of the Spaniards. The French knew the really hot varieties as 'cayenne'. In Mexico and North America the name 'chili' is used, and in central Europe both plants and fruits are known as 'paprika'.

Both annuals and perennials are included in the genus *Capsicum*, and their stems are more or less branched according to the species concerned. A section of stem will normally end in a flower, and then its side-shoots grow beyond it. Usually, the leaves are ovate to lanceolate in shape with a short point, and are shortly stalked. The small flowers are relatively inconspicuous but in structure resemble those of the potato. The superior ovary develops into a berry, which, in the sweet pepper, is divided into compartments in the lower half, and contains numerous seeds. In this case the fleshy wall of the fruit is usually quite thick, but in the hot, spicy forms it is noticeably thin. The shape and colour of the fruit can vary considerably. The colour can range from green, through yellow and bright red to deep purple, while the shape can vary from a slender, cigar-shaped 'pod' to roundish, lantern or balloon-shaped berries which taper at the end.

In the hollow interior of the *Capsicum* berry are numerous seeds, and both these and the adjacent tissue contain the alkaloid capsaicin. In 'paprika', this amounts to 0.3–0.5%, but in the extremely hot 'cayenne' there is 0.6–0.9% and even a minute quantity of this can be detected by the tongue. In addition the fruits contain vitamin C, vitamin A, and various colouring substances including capsanthin. In the sweet pepper, the alkaloid capsaicin is almost entirely absent.

From a botanical point of view, classification and circumscription of the species is particularly confusing, since the plant has been in cultivation for a long time and has been hybridised again and again. The most important species is *C. annuum*, whose natural distribution extends from southern North America, through Central America and the West Indies and north-western South America to north of Lake Titicaca. Another important species is *C. baccatum*, which occurs naturally in southern Brazil, northern Argentina and Uruguay. Photo 213 shows a cultivated form of *C. annuum*, and Photo 214 one of *C. baccatum*. *C. frutescens*, which grows up to 2m high and becomes slightly woody, is an entirely tropical species from the Amazon region. Linnaeus described it from plants which had been collected in India. It was introduced into that country shortly after its discovery in America.

Nowadays paprika is cultivated not only in the tropics but also in subtropical and other warm regions, e.g. Hungary, Italy, Turkey, the Balkans, and Spain, and the centre of cultivation of the hot, spicy 'chillies' has moved from the New World to Asia and Africa, as the plant is now in great demand by the inhabitants of these regions.

## Nutmeg

215

*Myristica fragrans* Hout.

**Nutmeg family** Myristicaceae

The use of nutmeg in China and India goes back to pre-Christian times. About AD 600 the first nutmegs were brought to Europe by the Arabs, but, because they were one of the most valuable spices, the Arab traders would not divulge their origin.

The nutmeg is native in the Moluccas and the Banda Islands, where it thrives in the hot, wet climate of the tropical rain-forests. It forms a tree up to 30m high with dark green, leathery leaves, male and female flowers being borne on separate trees. In cultivation the trees are kept down to a more convenient height. Male trees produce small white flowers in the axils of oblong-ovate leaves. The inflorescences of the female trees are composed of one to three flowers with a white, bell-shaped perianth and a one-celled ovary ending in a two-lobed stigma. The ovary develops into a fruit which resembles a peach and has a fleshy outer layer. It splits open along the front side and partially also at the back revealing a single seed, incorrectly called a 'nut'. The seed is surrounded by a woody, dark brown shell, which in turn is enclosed by a bright red aril with a lattice-like structure.

The aril loses its red colour as it dries, becoming a drab brownish yellow and hardening to a horny consistency. Like the seed, the aril is used as a spice and is known commercially as 'mace'. This has to be cracked open in order to obtain the seed, which is then dried to produce the 'nutmeg' of commerce. The seed contains 30% oil, 30% carbohydrate, and 7–16% of ethereal oils of which the most powerful is myristicin. This substance is by no means harmless, 4g being sufficient to cause human poisoning. Small and damaged seeds are heated and pressed to extract an oil known as 'nutmeg butter' which is used in ointments and perfumery. A mixture of oils called 'oil of nutmeg' is obtained by distillation and used in medicine externally and by the perfume industry.

## Vanilla

216

*Vanilla planifolia* Andr.

**Orchid family** Orchidaceae

Vanilla is the only economic plant amongst the 30,000 or so species in the family Orchidaceae. It is a climbing plant which occurs wild on forest-margins in Mexico, growing up to 10m high and supporting itself by means of tendril-like adventitious roots. The long, succulent, green stems bear sessile, thick, fleshy leaves which are lanceolate in shape. Racemes of attractive yellowish green flowers arise in the axils of the leaves (see illustration opposite). The flowers are only open for a few hours in the morning and fall off if not pollinated by humming-birds, particular kinds of insects or by hand. If fertilisation is successful, the inferior ovary, formed from three united carpels, develops within four weeks into a long, narrow capsule. This needs a further five to seven months to ripen properly.

If the fruits are to be used for flavouring, they must be gathered while they are greenish yellow and before they split open. They are first treated with steam or boiling water and then allowed to ferment. For this, the fruits are heated by being spread out on blankets in the sun for several hours, and then placed in airtight containers over night where they sweat. During this process, which lasts for four weeks, vanillin is

Vanilla (*Vanilla planifolia*). *Left:* portion of stem with inflorescence. *Right:* fruits.

produced, sometimes appearing as crystals on the outside of the fruits and turning them brown.

The cured fruits, often incorrectly called 'pods', contain 3.2–3.7% vanillin and 35 other aromatic substances. Until 1846 Mexico had the monopoly of the vanilla trade, but even before this, in 1819, the Dutch introduced the plant to Java and, shortly afterwards, the French brought it to Réunion (Bourbon). As the natural pollinators did not exist there, an alternative method was required. In 1841 a system of hand-pollination was developed which resulted in a larger fruit set. Madagascar is the largest vanilla producer, growing about two-thirds of the world crop.

Two other species, *V. pompona* from Venezuela and *V. tahitensis* from Tahiti, are found wild and also cultivated, but do not supply such a high quality product.

## Ginger 217

*Zingiber officinale* Roscoe

**Ginger family** Zingiberaceae

Ginger was already being grown in S.E. Asia in pre-Christian times as a valuable spice plant. Its home is on the Pacific Islands although it no longer occurs there wild. The part used is the tuberous rhizome, which can reach a length of 50cm, often branching like a human hand. The rhizome sends up sterile stems up to 1m high, bearing linear-lanceolate, pointed leaves, and also leafless, fertile stems up to 30cm high and thickened at the top like a club. From this enlarged top arise thick, fleshy bracts. In the axils of these bracts are the relatively small, yellow flowers with a purple lip, which only rarely produce seeds.

The rhizomes are plunged into boiling water, and then, if the outer skin is peeled or scraped off they are referred to as 'uncoated' ginger or, if left on, as 'coated' ginger. Both forms contain 0.6–3.3% of ethereal oils, also the pungent substance zingerone and a mixture of resins. The spice reached Europe in ancient times through Arab traders. Nowadays it is used predominantly in eastern Asia, and in India it is in almost daily use as an aid to digestion. Ginger has been of medicinal value since olden times. It is powdered and used to spice sausages, and is also employed in the manufacture of refreshing drinks such as ginger beer. The main areas of production are in China, India, Japan, the West Indies and Africa, with cultivation concentrated in southern China, Indomalaya and Jamaica.

## Annatto **218, 219**

*Bixa orellana* L.

**Bixa family** Bixaceae

The annatto occurs in the wild in the forests of Costa Rica up to an altitude of 800m, and is the sole representative of its family. From this single species have arisen all the cultivated forms which are spread nowadays throughout the tropics. The annatto can reach a height of 10m, but is generally kept lower, especially when it is being grown for the dye in its seed-coat. The large, long-stalked leaves are narrowly heart-shaped, and the lovely flowers of a delicate pink colour are arranged in panicles at the ends of the branches (Photo 218). The ovary develops into a slightly flattened capsule covered with soft prickles (Photo 219), and attached to its inside walls are 20–50 seeds, the outer covering of which contains an orange-red dye called bixin. Commercially, this dye goes under the name of annatto, orlean or roucou.

The colouring substance, which is completely harmless, is obtained by putting the seeds into hot water and collecting the reddish pigment which settles to the bottom. The dye is used in a variety of ways for colouring food, e.g. rice, cheese, margarine, butter and bakery products, also in cosmetics. Even today the plant plays an important role in the daily life of the Indians in the forest regions of the Amazon and Orinoco. The women use the dye in the seed-coat to beautify themselves and the men rub bixin paste on to their bodies before they go hunting. Their skin is lacking in pigment so they can probably protect themselves from the burning rays of the sun in this way. Because of its large leaves the annatto is also a favourite shade-tree, and it can be found whenever there is human habitation.

# Vegetables and Fruit

## Okra, Lady's Fingers, Gumbo 220

*Abelmoschus esculentus* (L.) Moench

**Mallow family** Malvaceae

The genus *Abelmoschus* comprises 15 species native in subtropical and tropical regions, but the only species of economic value is the okra which originates in tropical Asia. This is a bushy, herbaceous annual which grows up to 2.5m high and becomes woody at the base. It has long-stalked, lobed leaves with toothed margins. From the end of the stem arise up to 40 yellow, five-petalled flowers. After self-pollination these develop into slender, six-sided, beak-like fruits, 12–15cm in length, attached to the enlarged receptacle. The fruits are gathered while they are still unripe and tender, and provide a mild-tasting vegetable when cooked. Because they are very mucilaginous, they are particularly suitable for those suffering from stomach trouble, and if served with salt, pepper, oil and lemon juice they make a tasty salad. The seeds are eaten when roasted. They contain up to 25% of oil, which is extracted from the ripe fruits and used in the manufacture of margarine.

## Snake Gourd 221

*Trichosanthes cucumerina* L.

**Gourd family** Cucurbitaceae

The genus *Trichosanthes* comprises 25 species native in the Indomalayan archipelago and Australia. Its name is derived from the Greek words 'thrix', 'trichos' (hair) and 'anthos' (flower) and refers to the fringed corolla. The snake gourd is native in south-east Asia and is distinguished by its bright orange-red fruits. It is a quick-growing plant with strong, angular climbing stems 5–6m high. The leaves are more or less digitate with three to five lobes which are either entire or have rounded teeth. The flowers are white and the edge of the corolla is finely cut into narrow strips. The shoot-system of the plant is complex in form and the flowers are unisexual, the male being arranged in racemes, and the female appearing singly on the stems. The inferior ovary develops into a slender, highly decorative, berry-like fruit, whereas most of the other species in the genus have rounder, less strikingly coloured fruits. The plant known as *T. anguina* is probably only a cultivated form of *T. cucumerina*. It flowers at night and has snake-like fruits over 1m in length. The species illustrated is of more value as an ornamental than as an economic plant.

## Chayote, Christophine 222

*Sechium edule* (Jacq.) SW.

**Gourd family** Cucurbitaceae

The genus *Sechium* has a single species, native in Brazil. The origin of the generic name is unknown and the plant itself is no longer found in the wild. It used to be cultivated by the Aztecs but nowadays it is grown mainly in the West Indies and West Africa. The stems grow to more than 10m high, climbing by means of tendrils. Unisexual flowers develop in the leaf-axils, groups of male flowers appearing first, followed by individual female flowers. All the flowers are whitish in colour and very small. As well as tendrils, the plant has broadly ovate or triangular-ovate leaves, 10–20cm long, with entire margins. The ovary of the female flower develops into a pear-shaped, berry-like fruit, 8–15cm long and weighing up to 1kg. It is wrinkled at the base, covered with soft prickles, and contains only one seed. The hard, ovoid seed is up to 10cm in length and germinates inside the fruit. The fruit contains 7.5% carbohydrate, 1% protein and vitamin C.

The root of the plant becomes turnip-shaped, and forms new buds on its enlarged, disc-like top. These later grow into new shoots. The swollen main root produces lateral roots which form tubers at the end and can weigh up to 10kg. As these contain 20% starch they are carefully gathered, leaving the main root intact, and cooked like potatoes. The fruits can be used as a vegetable, often with a meat or fish filling. They can also be preserved or used as a high quality pig food. The young shoots and roots can be eaten like asparagus. The flowers provide food for bees, and the long stems are a source of fibre which is used in the making of hats, sacks and mats.

## Bamboo **223**

*Dendrocalamus* spp.

**Grass family** Gramineae

The genus *Dendrocalamus* comprises 25 species with a natural distribution extending from India to China. The young shoots of *D. hamiltonii* are a favourite vegetable while *D. giganteus*, the giant bamboo, is more a decorative plant. It is found mainly in damp places, usually near streams and rivers, and has a large, underground rootstock which sends up stems to a height of as much as 30m. At first the young shoots are sharply pointed and their base is enclosed by rudimentary leaves, regarded morphologically as leaf-bases. The actual leaf-blade is in the form of a slender point. A noticeable feature of the stem is its sharp differentiation into nodes and internodes. Its surface is glassy and extremely hard, and is more often yellowish brown or yellowish green than uniformly brown. The outer cell-layers of the stem are particularly rich in silicic acid. In humid localities the internodes are often covered with lichens, and rings of adventitious roots grow out from the nodes, and hang down almost like beards. The strong tendency to produce adventitious roots is due to the fact that the main root dies early.

The speed of growth in species of *Dendrocalamus* is very high, the stems growing 30–50cm in a day. In most species of bamboo, which can be separated from the Gramineae to form their own family, the Bambusaceae, flowers are produced only rarely, but when this happens, all the plants of one species flower at the same time. The parts above ground then wither and die, but the plant may regenerate slowly from the subterranean rootstock. The fruits (caryopses) which are formed contain a considerable amount of starch and are gathered like rice. A heavy fruit set is often followed by a plague of mice and rats.

Because a number of the larger species of bamboo flower very rarely, their systematic position is in many cases still uncertain. Some species have been found to flower every 32 years, but in species of *Sasa* the interval has been observed to be 60 years, while for members of other genera, there is an interval of 100 years between one flowering and the next!

## Avocado

224, 225, 257

*Persea americana* Mill.

**Laurel family** Lauraceae

The home of the avocado is in the mountain forests of tropical America. Even today most of the 150 species of the genus *Persea* are found in the tropics of the New World. The name 'avocado' is derived from the Aztec word 'ahuacatl', and the plant itself was in cultivation 8000 years ago, as has been demonstrated by relics from tombs in Central America. In 1519, the Spaniard Enciso came across the buttery fruit on the coast of Colombia, but it was the middle of the 19th century before it reached Asia.

The tree reaches a height of up to 20m, and has evergreen, elliptic-lanceolate leaves with prominent veins. The small, yellowish green flowers are produced in large numbers in racemes at the ends of the stems (Photo 224), but in spite of the abundance of flowers it is remarkable that only one in 5000 results in a fruit!

After fertilisation the pendulous, pear-shaped, green to reddish brown fruit begin to appear. They vary in size from only a few centimetres to 25cm in length according to variety (Photo 225). The fruit is botanically a berry, with a smooth outer skin (exocarp), a layer of cream to yellowish green flesh (mesocarp) that is as soft as butter when ripe, and, in the centre, a large, round, light brown 'stone'. The flesh has a nutty taste and contains 25% fat, a remarkably high amount, which contributes to the consistency and the special taste of the fruit. It is prepared in various ways as a salad by adding pepper, salt, oil and vinegar, but can also be cut in half, sprinkled with sugar and lemon, and then eaten with a spoon.

In all there are about 400 varieties belonging to three ecological races: the Mexican highland race with small, thin-skinned fruits, the Guatemalan race, which has long-stalked fruits with warty skins, and the West Indian lowland race with smooth, tough-skinned fruits. All three races are noted for their tolerance of cold and the aniseed-like smell of their leaves. The total production of avocado fruits is difficult to ascertain, but the largest producer-countries are Mexico, Brazil, Dominican Republic, Peru and Colombia.

## Banana, Plantain

226, 227, 228, 257

*Musa* spp.

**Banana family** Musaceae

The banana originates in the Indomalayan area where two species, *M. acuminata* and *M. balbisiana* occur wild. By hybridisation and domestication of these species, present-day bananas, usually eaten raw, and plantains, often cooked or made into flour, have been produced. In ancient times, the banana had spread from island to island in the archipelago, but it was probably Arab traders who brought the plant to the African continent. About 1510, Portuguese seafarers introduced it to the Canary Islands, and from there it reached Santo Domingo and shortly afterwards South America. Here it was spread so rapidly by the Indians, presumably fleeing from the advancing European conquerors, that at first it was thought that it originated in the New World. About the same time, the fruit became known in Europe also. Linnaeus named the banana which was rich in sugar *M. sapientium* and the one which contained abundant starch he called *M. paradisiaca* (Photo 227). Both are triploids obtained by crossing the wild species already mentioned so that the names given by Linnaeus are scientifically not tenable. The dwarf banana, previously known as *M. cavendishii*, is presumably a descendent of

*M. acuminata*. The word 'banana' is thought to be derived from the name for the plant in one of the Bantu languages of West Africa.

The banana is not a tree, but a herbaceous perennial 5–9m in height. It has a tuberous subterranean rhizome, from which arise impressively large leaves, in the case of *M. ingens* up to 15m high. The leaf-sheaths, or lower parts of the leaves, are folded within each other producing a 'false stem' from which the long, narrow blades protrude and spread out. Due to the strength of the wind, the blades often tear into strips between the lateral veins which run out at right-angles from the midrib (Photo 226). They then offer less resistance to the wind and tropical rain. In the centre of the folded leaf-sheaths, at ground level, a growing point forms on the top of the rhizome. During its nine or ten month period of development, it grows up inside the 'false stem' emerging at the top to produce a curved, overhanging inflorescence with a succession of reddish brown bracts covered with a bluish bloom. These unfold from the base to the tip and later fall off. In the axils of the lower 10–12 bracts arise 14–18 female flowers in double rows. Their ovaries develop into fruits without having to be fertilised, a process known as parthenocarpy. The next ten bracts contain bisexual three-petalled flowers, which are rich in nectar but do not develop further. In the axils of the upper bracts only male flowers are formed.

The fruits take three months to ripen, and most varieties contain no seeds but merely the black remains of the ovules. Only Fe'i bananas have seeds and erect fruits. Banana fruits are five-sided berries formed from three carpels, and the fleshy inner tissue is called pulp. The outer skin is formed from the receptacle as well as the carpels. It contains fibres and tannin canals which also produce drops of latex. As the fruit ripens, the innermost layers of the skin develop into a soft, fibrous tissue, allowing the skin to detach itself easily from the pulp. At first, both the banana and the plantain are rich in starch, firm in texture, and astringent in taste. But as they ripen, the starch is converted into sugar, and the fruits become soft and sweet. One of the principal aromatic substances present is isovalerianic acid.

After the fruit has ripened, the upper part of the plant withers and dies, but even before flowering-time, suckers are formed on the part of the stem above ground, and so the plant can propagate itself vegetatively. New plantings can be made with suckers that are four to seven months old and ready to flower. They are cut down to a height of 30cm and then separated from the mother-plant. After transplanting they quickly root and form new plants, the strongest of which are kept. Newly established banana plantations will yield fruit for 5–20 years.

The banana requires a warm, damp climate with an average annual temperature of 25°C and a rainfall of 1200–2000mm. Average annual yields vary from 30–60 tons per hectare, but can rise to 100–130 tons. The most important exporting country is Ecuador, followed by Honduras, Costa Rica and Panama. Apart from the consumption of bananas as raw fruit, they can be peeled when fully ripe and heated at a temperature of 60°C to produce dried bananas, a favourite in the USA. When bananas are ripe but still green, they can be made into a flour, to which water and milk are added to form a dietary food.

The banana has, in the course of time, become one of the most important tropical fruits. It has been in cultivation for a very long time and so there is a wide range of varieties which have resulted from the crossing of *M. acuminata* and *M. balbisiana*. The best and most widely cultivated in 'Gros Michel', but because it is susceptible to banana wilt, it is being replaced again and again by varieties such as 'Giant Cavendish', 'Robusta', or 'Lacatan'. A great favourite is the dwarf banana (Photo 228) which comes from China, but its fruits are less suitable for transport because of their thin skin. It is cultivated in the Canary Islands, and other places under the name 'Dwarf

Cavendish'. The variety 'Huamoa', grown in Hawaii, belongs to the same race. Another species is *M. basjoo* which grows up to 4m high and originates in southern Japan. It is a source of fibre. Another important fibre-plant is *M. textilis*, Manila hemp, from the Philippines, which has red or pale pink bracts. The Abyssinian banana is native in the open mountain forests of that country and grows there up to an altitude of 2000m.

**Production of Bananas**
(in thousand tons)

| *Country* | *1975* | *1978* | *1980* |
|---|---|---|---|
| Brazil | 5311 | 6176 | 6773 |
| India | 3633 | 3853 | 4500 |
| Indonesia | 3000 | 1764 | 1606 |
| Ecuador | 2544 | 2375 | 2073 |
| Philippines | 1423 | 2435 | 3800 |
| Thailand | 1382 | 2000 | 2000 |
| Mexico | 1194 | 1137 | 1515 |
| Costa Rica | 1121 | 1170 | 1187 |
| Colombia | 1050 | 1500 | 1200 |
| Panama | 989 | 742 | 1050 |
| Burundi | 897 | 950 | 972 |
| Venezuela | 860 | 1080 | 985 |
| Honduras | 852 | 1338 | 1330 |
| Papua-New Guinea | 840 | 885 | 898 |
| Tanzania | 750 | 803 | 780 |
| Bangladesh | 578 | 590 | 605 |
| Guatemala | 520 | 566 | 560 |
| Vietnam | 490 | 520 | 540 |
| Spain | 490 | 390 | 464 |
| Other countries | 6285 | 6618 | 6416 |
| Total | 34209 | 36892 | 39254 |

## Pineapple

**229, 230, 231, 257**

*Ananas sativus* (Lindl.) Schult.
Syn.: *A. comosus* (L.) Merr.

**Pineapple family** Bromeliaceae

The pineapple originated in the tropical regions of Brazil. It has been in cultivation since ancient times and numerous forms were being grown and selected by various Indian tribes long before the arrival of the Spanish and Portuguese in the New World tropics. On 4th November 1493 Columbus was handed some pineapple fruits by Indians on the Island of Guadaloupe, after he had accomplished his second crossing of the Atlantic. So he became the first European to have knowledge of the plant. In later years, many Spanish and Portuguese explorers reported the incomparable flavour of the pineapple, and, in a letter dated 1513 to King Ferdinand of Spain, Oviedo made the first drawing of the fruit. Soon afterwards it reached the Old World, for in 1548 it was being planted in Madagascar and in 1590 in India. Since then it has spread over the whole of the tropics.

The pineapple is a perennial with a short stem and usually spiny-edged leaves, 30–100cm long, arranged in a rosette (Photo 231). Offshoots with small rosettes of leaves arise in the axils of the large leaves and serve to propagate the plant vegetatively. After one or two years, the stem lengthens and forms a spike-like inflorescence at the end which is up to about 15cm long and has a thickened axis. It consists of numerous, pink-coloured, long-pointed bracts with pink three-petalled flowers in their axils (Photo 229). The flowers turn into fruits without being pollinated, and the inferior ovaries develop into berries, which, together with the axis of the infloresence and the bracts, form a compound fruit or syncarp. Only the roughly diamond-shaped and flattened sides of the individual fruits can be seen, making up the surface of the aggregate fruit. The upper bracts of the infloresence do not have flowers in their axils, but turn green and leaf-like. They form a tuft on top of the fruit which can be cut off and used for vegetative propagation (Photo 230). Both the Spanish name 'piña' and the English name 'pineapple' were given to the fruit because of its resemblance to a pine cone.

The pineapple is related to bromeliads used as house-plants, and is a typical tropical plant in that it thrives best with an average annual temperature of 25°–32°C, a rainfall of 1000–1500mm and high humidity. There are three principal cultivated varities. Var. *sativus* is seedless, var. *comosus* produces fertile seeds; and var. *lucidus* has the advantage of having leaves without spines. Various Indian tribes in South America have obtained, quite independently of each other, a number of spineless forms by a

**Production of Pineapples**
(in thousand tons)

| *Country* | *1975* | *1978* | *1980* |
|---|---|---|---|
| China | 859 | 916 | 888 |
| U.S.A. | 617 | 635 | 622 |
| Brazil | 515 | 569 | 572 |
| Thailand | 500 | 1250 | 1500 |
| Philippines | 360 | 500 | 600 |
| Mexico | 262 | 300 | 583 |
| Malaysia | 245 | 197 | 203 |
| Ivory Coast | 233 | 300 | 320 |
| South Africa | 184 | 185 | 208 |
| Bangladesh | 124 | 142 | 142 |
| Zaire | — | 165 | 156 |
| Other countries | 1458 | 1677 | 1842 |
| Total | 5357 | 6836 | 7636 |

process of selection. As a rule, a pineapple plant yields two crops, the first after 15–24 months, and the second, from side-shoots, after a further 15–18 months. Details of production from various countries are given in the table on p. 218. The edible part of the fruit contains vitamins $B_1$, $B_2$, C and the enzyme bromelin, which breaks down proteins and also promotes digestion, a fact already recognised by the Indians. World production of pineapples doubled between the years 1948 and 1965, and has further increased since then. Besides being grown for its fruit, the pineapple has for centuries been cultivated as a fibre plant. The fibre is obtained from the leaves by a lengthy manual process consisting of steeping and bleaching. It is then used to make the famous mantillas and other articles suitable for the hot tropical climate.

## Passion fruit, Purple Granadilla 232, 234

*Passiflora edulis* Sims

**Passion-flower family** Passifloraceae

The genus *Passiflora* comprises 400 different species, the majority of which originate in the tropics of the New World. Only a few species come from Asia, Australia and Polynesia. Most of them grow in forest regions although there are some in Central America which are found on the Pacific coast. The most important species economically is *P. edulis*, known in its native country of Brazil under the names 'maracuja' or 'granadilla', which is a perennial climber with long stems, attaching itself to its supports by means of axillary tendrils. It has alternate, three-lobed leaves, in the axils of which solitary flowers appear whose structure is similar to that of *P. caerulea*, already described on p. 55. The fruit is a berry, the size of a hen's egg (Photo 234), whose outer skin is at first juicy but acquires a parchment-like quality as it ripens. Inside is a mass of pulpy tissue, containing numerous seeds, which has a pleasantly aromatic, somewhat acid taste, and is usually spooned out when the fruit is eaten. A dull orange juice, also known as 'maracuja', is obtained from the ripe fruit. At maturity the outside of the fruit is normally purplish red to deep purple, although there is a yellow-skinned mutant. The passion fruit is increasingly being grown in plantations, and is now found throughout the tropics. The plants often intertwine, forming huge canopies, and the fruits hang down in the shade of the leaves. A single plant will produce about 100 fruits in a year. The alkaloids present in the leaves have the effect of reducing blood-pressure.

## Giant Granadilla 233

*Passiflora quadrangularis* L.

**Passion-flower family** Passifloraceae

The giant granadilla is the most striking of the economic plants in the genus *Passiflora* and is more tropical in its requirements than *P. edulis*. The name of the species refers to its square stem (quadrangularis = four-angled) which bears large, undivided leaves with a heart-shaped base and a pair of prominent stipules. The stalks of the leaves usually have six glands. The large flowers are highly ornamental on account of the radiating series of white and purple filaments. The fruits are particularly large, reaching a size of about 25 × 18cm and a weight of 4–5kg. The inside of the melon-like fruit is watery, and the reddish juice has a slightly acid taste. The pulp, including the seeds, is eaten, after sweetening. The name 'granadilla', which is applied to all kinds of passion-fruit, refers to their similarity to the pomegranate. Besides the giant granadilla, the species *P. alata* and *P. macrocarpa* are also cultivated for their palatable fruits. The fruits of *P. ligularis*, *P. laurifolia*, *P. mollissima*, and *P. mixta* are also edible, but, like *P. edulis*, only the pulp is eaten, as the rind becomes dry and leathery. Many species of *Passiflora* form underground tubers from which long climbing stems arise.

## Mango

**235, 236, 237, 257**

*Mangifera indica* L.

**Cashew family** Anacardiaceae

Next to the banana, the mango is the most important fruit in the tropics, and is comparable with the apple in temperate regions. Its home is in the mountains of central Burma and in the foothills of the Himalaya in eastern India. Domestication of the tree took place very early in India, for the mango was mentioned as a cultivated plant in Sanskrit writings of 4000 years ago, and by 500 BC it had spread from India to Malaysia and the eastern asiatic islands. During the time of the voyages of discovery Portuguese seafarers brought the tree to West Africa and Brazil, and, quite independently, Spanish ships transported mango seedlings from the Philippines to Mexico. The plant first reached the West Indian islands about 1740. In keeping with its origin, the mango tree requires an average annual temperature of at least 19°C and 1000mm rainfall. It has recently been planted in Egypt, Israel and South Africa.

At maturity, this impressive tree can be 30m high (Photo 235) with a dense crown of evergreen leaves, making it useful not only as a fruit tree but as a shade tree also (Photo 237). In India, some trees even reach a height of 40m and some of their roots penetrate the ground to a depth of 7–8m. However, the root system also lies near the surface, spreading out 10m from the trunk, the same distance as the crown. As with many tropical trees, the leaves are produced in flushes, appearing in large clusters at the same time, and hanging limply down from the ends of the branches. The leaves are alternate, lanceolate, up to 30cm long and 7cm broad, and have stalks 5cm in length. When young the leaves are a delicate reddish colour, due to the formation of anthocyanin. Later they turn dark green and leathery, with a pale green midrib.

The inflorescence is a panicle, composed of 2000–5000 small, pale greenish yellow flowers. Some of the flowers are purely male, the others are bisexual. The five sepals are 4mm long and hairy on the outside, the five petals are 3–6mm in length. Only one of the five stamens is fertile, which accounts for the relatively small number of fruits developing from the superior ovaries. The flower has an enlarged, disc-shaped base, known as the receptacle, which is a characteristic of the Anacardiaceae.

Pollination is carried out by flies rather than bees. Other reasons for reduced fruit-production are the need for the right kind of soil and the fact that many of the fruits drop during the course of development. In a number of Indian varieties, only one in 150 fruits reaches maturity.

While the turpentine content of the skin of the fruit is variable, the mango, called 'manga' in Brazil, is one of the most popular tropical fruits. The fruit is termed botanically a drupe and is rather pear-shaped or kidney-shaped in form (Photo 236). As a rule, the fairly thick outer skin is yellowish green or reddish in colour, and the juicy golden yellow flesh is more or less fibrous, the fibres extending into the flattened stone. The size of the fruits varies considerably. On an average they are 8–12cm long, but occasionally they may exceed 20cm and weigh over 3kg. Numerous forms and races have arisen in the course of time, and these vary greatly in the amount of turpentine, sugar and vitamins they contain and consequently there are differences in taste.

The fruit is extremely rich in Vitamins A and C. There is a high percentage of fructose, and protein and fibre are also present. In India, the mango is of particular importance as food for the population. The fruits are usually eaten raw, often while still unripe, but they are also used in the manufacture of preserves, juice and pickles. When dried and ground they form a powder called 'amchur' which is added to soups and chutneys. Export of fruits is on a relatively small scale because they deteriorate rapidly, but if transport and storage conditions were to improve, consumption in countries outside the tropics would undoubtedly increase.

**Production of Mangoes**
(in thousand tons)

| *Country* | *1972* |
|---|---|
| India | 8400 |
| Brazil | 665 |
| Pakistan | 660 |
| Bangladesh | 480 |
| Mexico | 220 |
| Dominican Republic | 179 |
| Tanzania | 150 |
| Philippines | 135 |
| Colombia | 105 |
| Honduras | 90 |
| Other countries | 616 |
| Total | 11700 |

## Pawpaw, Papaya

238, 239, 240

*Carica papaya* L.

**Pawpaw family** Caricaceae

The pawpaw was being cultivated by the Indians of Central America and Brazil in pre-Columbian times. They selected it from the numerous species of the genus *Carica* which grow in the hot, humid, lowland and coastal forests of this area. The Spaniards came upon the pawpaw in Panama and brought it to the West Indian islands. At the end of the 18th century it reached Asia and nowadays it is cultivated throughout all tropical regions.

It is a short-lived tree, up to 6m high, with a green, cabbage-like trunk, the smooth surface of which clearly bears the scars where leaves have fallen off. The leaves are large and long-stalked, with deeply divided, palmately-lobed blades, and are arranged in a tuft at the top of the trunk. The inflorescences arise in the axils of the leaves, those on male trees hanging in pendulous panicles (Photo 238), while those on female trees are in shortly branched clusters. The female flowers are somewhat larger and yellowish white, and the ovary is formed from three to five united carpels. On top of the ovary are five stigmas, each one broadened at the top and deeply cut into a number of segments. Occasionally bisexual flowers occur.

Fruits are produced the whole year through after pollination by insects. They are ovoid to pear-shaped, as large as a human head, and somewhat resemble melons (Photo 239). The fruits usually weigh 500–1000g although they may attain a weight of 5kg. Their yellowish green skin encloses the juicy pulp, which is pale yellow to orange-red in colour and has a buttery consistency. It is free from acid and has a delicate taste, somewhat similar to that of the musk melon. The surface of the central cavity in the fruit is covered with dark grey seeds, as large as peppercorns, which taste like cress.

The fleshy part of the fruit contains 5% carbohydrate and 0.6% protein, also abundant vitamin A and C. The protein-splitting enzyme papain is also present, and this helps in the digestion of meat. It is obtained from the abundant sap which oozes from cuts made in the bark and from immature fruits. The liquid is dried to form a greenish grey powder, commercially known as papain, which is used medicinally in cases of weak digestion and also as a meat-tenderiser. In addition, it is used in the manufacture of chewing-gum, and in the textile industry, where it helps to prevent shrinkage of wool and silk. The pawpaw is one of the most frequently cultivated fruit-trees in the tropics, and together with bananas and manioc, is often planted on small farms for the farmer's own use. Recently, the pawpaw has been grown in large plantations for export purposes, with yields of 30–50 tons per hectare.

## Sweet Sop, Sugar Apple
*Annona squamosa* L.

## Sour Sop
*Annona muricata* L.

## Bullock's Heart
**241, 242, 243, 257**

*Annona reticulata* L.

**Custard-apple family** Annonaceae

The genus *Annona* comprises some 80 species, most of them native in the New World tropics. Some members of the Annonaceae were prized for their fruit by the early inhabitants of South America, as has been established by relics found in graves in Peru. Every species has its own particular taste, which is not comparable to any fruit of the Old World.

The sweet sop is native in the Caribbean islands, but is cultivated throughout the tropics. The tree reaches a height of about 12m and has deciduous, elliptic-lanceolate leaves. It produces flowers similar to those of a *Magnolia*, but smaller.

The carpels develop into berries 9–15mm long, whose fleshy tissue coalesces to form a multiple fruit (Photo 241). Each individual fruit contains a shiny, black seed, and it can easily be separated from the others. The pulp is aromatic and has a sugary sweet taste.

Closely related to the sweet sop is the sour sop, the natural distribution of which can no longer be ascertained. The tree grows up to 12m high and its fruit weighs up to 2kg. In this species the individual fruits have grown together up to their tips, giving the impression of a single large fruit (Photo 242). The leathery green rind is smooth, apart from the fleshy spines which represent the tips of the individual fruits. The pulp has a woolly, almost fibrous texture. It is strongly aromatic and has an acid taste, making it very suitable for refreshing drinks.

The bullock's heart, *A. reticulata* (Photo 243), is a tree native in Central America, which grows up to 8m high and has long, narrow, evergreen leaves. The individual fruits form a homogeneous structure, whose appearance gave it its common name. The only outward sign that it is a compound fruit is the network of lines, indicating the edges of the individual fruits, which is just visible on its surface. The pulp is aromatic, with a sweet, creamy taste, and, as in all species of *Annona*, encloses large, shiny seeds. Like the sour sop, the asymmetrical form of the fruit is due to fewer carpels on one side being fertilised and therefore remaining undeveloped.

According to gourmets the cherimoya, *A. cherimola,* is the best tropical fruit. This delicate little tree, 5–8m in height, cannot tolerate the hot, lowland climate, and only thrives at altitudes of 800m or more in the northern Andean countries. Even during the time of the Incas, the cherimoya was known as a fruit tree. Its fruits are also heart-shaped, 10–15cm in diameter, and regularly patterned by grey-green scales. The pulp is juicy and sometimes granular in texture.

## Tamarind **244, 245**

*Tamarindus indica* L.

**Senna family** Caesalpiniaceae

The tamarind is native in Africa, although the generic name is derived from the Arabic 'tamar-Hindi', Indian date. The genus contains only one species, which is found nowadays in all the tropical regions of the earth. The tamarind grows up to 25m high and its trunk can be up to 8m in circumference. Its branches resemble those of a *Gleditsia* but they are without spines. The evergreen leaves are pinnate, and are made up of 12–15 pairs of leaflets with very small, deciduous stipules at the base of the leaf-stalk. The beautiful pale yellow flowers, streaked with red, are in racemes at the ends of the branches. The single carpel develops into a rounded, light brown pod, which remains closed (Photo 244). The seeds are surrounded by a sticky, brown pulp, which was commonly used in former times under the name 'pulpa tamarindorum' as a gentle laxative. It serves both as a food and, especially in Central America, as a refreshing acid drink. In India, the pulp is eaten raw sweetened with sugar. It is also used in the manufacture of syrup and sweets. The tamarind is very popular as a shade-tree, since its delicate, feathery foliage softens the glare of the sun in a very pleasant way. Its timber is dark, hard, and durable, and is used to make all kinds of implements.

## Guava **246**

*Psidium guajava* L.

**Myrtle family** Myrtaceae

The genus *Psidium* comprises some 100 species, whose distribution extends from Uruguay, through South and Central America to Mexico and the West Indies. The most important species, as a producer of fruit, is *P. guajava*, the guava. It forms a small, rather gnarled tree up to 10m high with smooth, light brown bark, which peels off in small flakes. The young side-shoots are four-angled. The leaves are in pairs, oblong-elliptic to ovate, and the underside is covered with fine hairs. The lateral veins are clearly visible, slightly sunken on the upper side, and prominent below. The white flowers, 2.5cm across, have numerous stamens and appear singly or in groups of two or three in the axils of the leaves. The inferior ovary develops into a globular to pear-shaped berry with a smooth skin and the persistent calyx at the end (Photo 246). The interior of the fruit is composed of whitish, yellowish or pinkish tissue of a creamy consistency, and numerous hard, angular seeds. The flesh is very rich in vitamin C, but vitamin A, iron and phosphorus are also present in fairly high proportions.

The fruits are 4–10cm long and are used in the manufacture of aromatic jellies and jams. They are also eaten cooked. The trees begin to bear when they are two years old, and produce abundant fruits from their seventh or eighth year for the following 30 years or so. Yields amount to 30–40kg for a single tree. In view of increasing demand, especially for its juice, cultivation of the guava has increased throughout the whole tropical region, particularly in India, where seedless forms are known. Besides *P. guajava*, other species have been introduced into other continents from South America.

## Cashew **247**

*Anacardium occidentale* L.

**Cashew family** Anacardiaceae

The cashew is native in the relatively dry areas of the Caribbean region and north-eastern Brazil, and its English name is derived from the Indian name 'kaju'. In the coastal plains of Venezuela and north-eastern Brazil, the plant is called 'marañón'. As early as the 16th century it was brought by Spanish seafarers from its original home to India. On good soils the cashew tree can reach a height of 12m, but on poor soils it is often only a shrub. Its leaves are alternate, ovate, 15–20cm long, prominently veined in pale green, and of a leathery texture. The flowers are in panicles at the ends of the branches and may be purely male or bisexual. Only a few flowers in the panicle develop into fruits. These are kidney-shaped and are attached to the fleshy, swollen fruit-stalk. This shiny, red stalk is known as the 'cashew-apple', while the true fruit or nut hangs from its enlarged end.

Cashew apples and nuts stand in third place in world trade, after bananas and pineapples. The fleshy stalk is highly regarded, especially in dry, hot regions, as a refreshing, slightly acid fruit. The juice can be consumed fresh or fermented and made into wine, and the pulp used in making jam. The true fruits are even more important. Nowadays, cashew-shell oil is used medicinally in various ways. The oil can also be distilled and used in the manufacture of varnishes, cements, inks etc. It is also used to prevent termites from attacking wood, and for brake-linings, as it can be polymerised to form a kind of rubber. Only after the oil has been extracted is the fruit opened by hand and the seed removed. This is then roasted to become the well-known 'cashew nut' which contains 45% oil and 20% protein. The seeds are sometimes crushed to extract this edible oil. Of the many countries producing cashew apples and nuts, India takes first place.

## Sapodilla

**248, 249**

*Manilkara zapota* (L.) Van Royen
Syn. *Achras zapota* L.

**Sapodilla family** Sapotaceae

The genus *Manilkara* comprises 80 different species distributed throughout the tropics. The most important economic plant in this genus is the sapodilla, which is cultivated nowadays throughout the tropical regions of the world. Its centre of cultivation, however, is still in the New World tropics. It is a tree, 10–15m high, with elliptic leaves in pairs or almost opposite each other in clusters at the ends of the branches (Photo 248). Six-petalled flowers arise in the axils of the leaves. The ovary, usually formed from six united carpels, develops into a berry with a green, grey or orange coloured skin. The outer surface can be smooth or slightly rough due to the formation of corky tissue (Photo 249). The size and shape of the fruit can vary considerably, but on average it is as large as an apple. The berries are called 'níspero' in Latin America and 'naseberry' in the West Indies, and the flesh of the fruit is soft and sweet, and has granules embedded in it. The seeds are up to 2cm long, flattened, and black when fully ripe. The tree is the source of a very hard and durable timber. In the trunk, as in all other parts of the plant, there are latex canals which produce a milky sap when the surface is scratched. The latex solidifies on being boiled and is then known as 'chicle'. This provides the basis for the manufacture of chewing gum. A tree 10–15m high will produce 15–18kg of dried latex annually if properly tapped. This represents an important source of income for Central American countries, since it can be sold in vast quantities, particularly to the USA. Quite independent of western civilisation, the Indians of the upper Orinoco use the dried latex for the same purpose.

## Genipa, Marmalade Box

**250, 251**

*Genipa americana* L.

**Madder family** Rubiaceae

The genus *Genipa* contains only four species, which are native in the West Indies and South America. The best known of these is the genipa or marmalade box which is now cultivated throughout the tropics. It has pairs of narrowly elliptic leaves and flowers which are twisted in bud. The ovary develops into a berry-like fruit about 10cm long and elliptic in longitudinal section. Surrounded by the flesh of the fruit, the rows of seeds lie embedded in a jelly-like pulp, which has given rise to the name 'marmalade box' used in some parts of its native region (Photo 250). Considering that the plant is a member of the Rubiaceae its fruit is very large. Only in tropical regions do representatives of this family produce fleshy fruits, and the genipa is an extreme example of this phenomenon.

## Langsat

**252**

*Lansium domesticum* Jack

**Mahogany family** Meliaceae

The langsat is a typical S.E. Asian fruit-tree, which, like the durian, rambutan, and mangosteen, only thrives in a humid, equatorial climate. It grows to 9–12m high and has dark green, pinnate leaves, each leaf consisting of five to seven elliptic-lanceolate, pointed leaflets, 12–25cm in length. Dense racemes or spikes of shortly stalked, five-petalled, sweetly scented flowers are produced on the larger branches. These develop into bunches of up to 20 oblong, straw-coloured fruits, each 4cm long. The berries are composed of five carpels, easily separable from each other, and containing a white, fleshy tissue. Each berry has one to three green, bitter tasting seeds. The flesh, on the other hand, is aromatic and has a pleasant, slightly acid teaste which is most refreshing. There is a particular variety of the langsat known as 'duku', which is distinguished from the slender langsat tree by its broader crown and denser foliage. The fruits are up to 6cm long, and there are 8–12 fruits in a bunch. The rind of the fruit is coarser than that of the langsat and, in contrast, contains no latex. Both kinds are found mainly in the Indomalayan archipelago.

## Malay Apple, Pomerac **253**

*Eugenia malaccensis* L.

**Myrtle family** Myrtaceae

The Malay apple is native in Malaysia and, like other members of the genus, is found in the rain-forests of the lower mountain region. The tree reaches a height of 15m and has glossy green, elliptic to lanceolate leaves, 20–35cm long. The branches bear dense clusters of conspicuous, four-petalled, bright pink flowers with numerous stamens. The two-celled ovary develops into a red fruit as large as an apple and smelling like a rose, which is eaten raw or made into preserves. Nowadays the Malay apple is cultivated in all tropical countries as an ornamental and a fruit-tree. In polynesian legends it is regarded as a sacred tree, which is why figures for temples are carved from its wood.

In addition to the Malay apple, the rose apple, *E. aquea*, which also originates in south-eastern Asia, is another important decorative and economic plant. This is a tree with a broad crown, which bears yellowish or white pear-shaped fruits with a taste resembling apricots. Its stems and leaves were formerly used as a dye. Of the numerous species of the genus *Eugenia*, which also occurs in South America, two in particular must be mentioned. The wax-like appearance of the pink or white, pear-shaped fruits of *E. samarangensis* give rise to its common name of wax apple, while the watery rose apple has pinkish red, flattened fruits which distinguish it from the two others. In contrast to related species which have a loose, spongy flesh, the fruits of the watery rose apple have a firm, rather brittle texture.

## Mangosteen **254**

*Garcinia mangostana* L.

**Mangosteen family** Clusiaceae

The mangosteen is another typical S.E. Asian fruit-tree. Its natural region of distribution is the tropical rain-forests on the eastern coast of Malaysia, though the plant is cultivated here and there in the entire equatorial belt. The tree grows up to 12m high and its arching branches, set low on the trunk, are densely leafy. The dark green leaves, more than 20cm long and up to 10cm broad, are arranged in pairs. The handsome flowers, 4–6cm across, are in simply branched inflorescences called dichasia. The flowers open in the afternoon, and the petals quickly fall while the sepals remain and are still visible as the fruit ripens. Almost all the flowers are purely female, and these develop into purplish brown fruits, up to 9cm in diameter, and somewhat resembling a tomato. At the tip are five to eight wide-spreading, slightly woody segments, which represent the remains of the stigmas. The fruit is a berry, and under the layer of dark red, slightly fibrous flesh, which is barely 1cm thick, there are five to eight fleshy, white segments (arils) having an extremely pleasant taste. Many people, in fact, consider the mangosteen as the choicest fruit of the tropics. As development of the ovules takes place without fertilisation and only a few sterile stamens are formed, the seed-coats frequently contain no seeds. They are either eaten fresh or can be cooked.

A tree does not begin to bear until it is 10–15 years old, but then it produces 500–1000 fruits annually. Other species of *Garcinia* are found in south-eastern Asia but none taste so fine as the mangosteen.

## Durian 255

*Durio zebethinus* Murr.

**Cotton-tree family** Bombacaceae

Because of its extraordinary fruits, the durian is one of the most interesting fruit-trees of the tropics. Its cultivation is limited to south-eastern Asia with centres in Thailand, Malaysia and Indonesia, where it has been grown for centuries. The durian is without any doubt the best known fruit of Malaysia. Its natural distribution lies in the western part of that country and in Borneo.

The tree grows up to 40m in height, with a slender trunk and a high crown of stout branches which spread out almost horizontally. By contrast, the tree is bushy in its young state with branches reaching almost to the ground. The entire, pointed leaves, lanceolate to elliptic in shape, are up to 28cm long and 10cm broad, and their colour varies from bronze to olive green. They have a smooth, shiny upper surface and the underside is covered with silvery grey hairs. The flowers, 5–6cm long, are arranged in clusters and smell like sour milk. They open about 3 p.m. and all the perianth-segments, as well as the stamens, fall off before the following morning.

The ovary develops into an elliptical fruit, weighing up to 3kg, and as large as a human head. Its surface is covered with thick spines, and is green at first but changes to greenish yellow when fully ripe. The fruit is a capsule, splitting into five parts at maturity (Photo 255), to liberate the rows of brown seeds as large as chestnuts. The seeds are surrounded by a soft, juicy, cream-coloured pulp, the aril. The pulp itself has a very pleasant taste, but the shell of the fruit emits a highly penetrating odour reminiscent of rotten onions. Those who are accustomed to the offensive smell prize the fruit greatly for the creamy texture of the aril.

For the inhabitants of south-eastern Asia, the eating of a durian fruit is a special delight, for the aril has a high nutritive value and is regarded as an aphrodisiac. The fruits are sold in the markets everywhere and are best eaten immediately after opening. The pulp decays rapidly on contact with the oxygen in the air and then acquires a sour taste. Because of its smell and taste, no other fruit has been so variously and inconsistently described as the durian. The aril is also eaten in a fermented state and even made into a kind of cake. The seeds are roasted, or cut into slices and cooked in oil. The tree begins to fruit when it is seven years old and produces two harvests a year, one from November to February and the main one from June to August. During those periods the air in the main centres of cultivation is filled with the smell of the fruits, simultaneously pleasing and repulsive.

## Rambutan

256

*Nephelium lappaceum* L.

**Soapberry family** Sapindaceae

The rambutan, like the durian, is a typical S.E. Asian fruit. Its home is in the west of Malaysia, where it grows wild in the tropical rain-forests even today. Cultivation of the plant is only possible in the hot and humid tropical areas round the equator. It is a decorative, bushy, broad-crowned tree up to 20m in height which has a strikingly dark brown bark with small markings. Its dark green leaves, consisting of two or three pairs of elliptic, slightly pointed leaflets, are somewhat glossy on the upper side. The leaflets on one side of the common stalk are not exactly opposite those on the other, and their size varies between 9–28cm in length and 5–15cm in breadth.

The inconspicuous, greenish flowers are in long-stalked panicles, and later develop into ovoid or globular fruits, 6cm long. These nuts have a thin parchment-like shell with a panelled surface, each panel with a long, soft, red spine at its centre. The outer skin changes colour from green to red as it ripens, though yellow-fruited races are known. Inside the fruit is a single seed, enclosed in a white, transparent covering which is sweet and juicy. This aril is not attached to the shell of the fruit. It has a refreshing sweet-acid taste and is rich in ascorbic acid (vitamin C). The fleshy aril is usually eaten raw, but may be stewed. The oily seeds are sometimes eaten roasted or are made into rambutan tallow. As a rule, the rambutan tree fruits twice a year. The main crop is produced from May to July, and a second, smaller crop in December.

In addition to the rambutan, people in the west of Malaysia and on Sumatra are familiar with the pulasan, *N. mutabile*, a lower-growing tree. This has leaves consisting of two to four pairs of leaflets which are narrower and lanceolate in shape. The surface of the fruit is covered with short blunt spines, in contrast to the long, soft spines of *N. lappaceum*, and it becomes darker as it ripens. The aril is white to yellowish and very sweet. The onset of ripening occurs somewhat earlier or later than in the rambutan. Different races have arisen within this species too. Another closely related species is the longan, *N. longana*, which is found wild in the tropical rain-forests of Malaysia, Sumatra, Borneo and Celebes. It forms a stately tree up to 20m in height, and its leaves, composed of three to five pairs of leaflets, vary considerably in shape and size. The very small, white, scented flowers develop into globular, hard-shelled fruits, 3cm in diameter. The outer skin is brown with lighter brown panelling. Inside is a single large seed enclosed in a relatively thin but sweet aril.

# Citrus Fruits

The genus *Citrus* which belongs to the family Rutaceae comprises 60 different species with a natural distribution in the Indomalayan region and in China. From this rich assortment, a number of species have been selected and crossed to form a group of plants of economic value which are cultivated not only in the tropics but in areas with a subtropical or mediterranean climate. Many of the original species come from higher altitudes within the tropical belt, where the climatic conditions can only be described as subtropical. This means that not all citrus fruits are tropical plants in the narrow sense. Most species are cultivated in subtropical regions, often in the intermediate zone between the tropics and subtropics. For this reason only two species have been illustrated, but at least these can be regarded as tropical if geographical degrees of latitude are taken as the criterion.

Almost all citrus plants are grown for their fruits which are a particular kind of berry. A berry is composed of three layers as follows: first there is the smooth external skin, known as the exocarp, which gradually merges into the mesocarp. This consists of an outer part, coloured orange or yellow by carotenoids. The inner part of the mesocarp is white, spongy when ripe, and contains numerous glands which produce ethereal oils. After the mesocarp comes the endocarp, a thin membrane surrounding the carpels or segments. Multicellular hairs grow out from the inner walls of the carpels and fill with juice to form pulp vesicles. They lie more or less close to each other, and enclose the seeds situated in the centre. In 'blood oranges' the cells forming the pulp contain the red, soluble pigment anthocyanin. 'Navel oranges' have a more complex structure in that there is a second, smaller ring of carpels above the main group. This smaller ring contains fewer carpels but these are usually well developed and appear as a small orange embedded at the apex of the normal fruit.

The most important citrus fruits are described in detail below.

Composition of citrus fruits per 100g of edible pulp (average values) (according to Souci et al.)

| *Constituents* | *Orange* | *Mandarin* | *Grapefruit* | *Lemon* |
|---|---|---|---|---|
| Water | 85.70g | 86.7g | 88.6g | 90.2g |
| Protein | 0.96 | 0.7 | 0.7 | 0.7 |
| Oil | 0.26 | 0.3 | 0.2 | 0.6 |
| Carbohydrates | 9.14 | 10.6 | 9.8 | 7.1 |
| Raw fibre | 0.53 | 1.0 | 0.3 | 0.9 |
| Mineral substances | 0.50 | 0.7 | 0.4 | 0.5 |
| *Vitamins* | | | | |
| Vitamin $B_1$ | 0.071mg | 0.06mg | 0.053mg | 0.051mg |
| Vitamin $B_2$ | 0.051 | 0.03 | 0.031 | 0.02 |
| Nicotinamide | 0.26 | 0.2 | 0.23 | 0.17 |
| Vitamin C | 51.0 | 30.0 | 45.0 | 53.0 |

## Sweet Orange

*Citrus sinensis* (L.) Osb.

**Rue family** Rutaceae

This is the best known of all citrus fruits. It originates from China and was being used there in 2000 BC. Long before the birth of Christ it had spread across India to Babylonia. In the baroque period in Europe oranges were being cultivated, by horticulturally minded princes, in glasshouses which were called orangeries. The fruit grown here, however, was mainly the sour or Seville orange. The first outdoor cultivation of oranges in Europe began only at the end of the 18th century, in Spain. There are now numerous varieties of the sweet orange, many with fancy names.

Production of oranges continues to expand and at present amounts to more than 30 million tons. They are an important source of nourishment, chiefly because of their refreshing acids and vitamins, vitamin C being by far the most abundant. The composition of the main citrus fruits is given in the table on p. 229. Botanically the orange is a tall shrub with somewhat spiny or sometimes spineless branches. As in most cultivated forms of citrus fruits, the leaf-stalk is slightly winged.

## Mandarin, Tangerine

*Citrus reticulata* Blanco
Syn.: *C. nobilis* Andr. non Lour.

**Rue family** Rutaceae

Like the sweet orange the mandarin has numerous varieties. One of these is known as the tangerine, but strictly speaking this should be called botanically *C. deliciosa* Ten. Of all the forms of the mandarin this has the smallest fruits, and these are usually canned for commercial use. The mandarin is a tropical plant and its home is in south-eastern Asia. The wild form contains numerous seeds. The seedless 'clementine' appeared as a mutant in 1912 at Oran in Algeria. 'Satsumas', named after the Japanese province of Satsuma are also usually seedless. Individual races of *Citrus* species can also be distinguished biochemically. The mandarin itself is a very spiny shrub or small tree. It produces small leaves, with a maximum length of only 3.5–4.0cm, which have narrowly winged stalks.

## Lemon

*Citrus limon* (L.) Burm. f.
Syn. *C. medica* L. var. *limon* (L.) Osb.

**Rue family** Rutaceae

The lemon was already known in China about 500 BC and reached Europe between AD 1000 and 1200. Nowadays it is cultivated in Asia, South and Central America, southern North America, and here and there in Europe too. Its yellow fruits are elliptic when cut lengthwise and have a distinct tip. They contain 3.5–7.0% citric acid, and are very rich in vitamin C. The lemon forms a tree 7–8m in height and its branches have stout, stiff thorns. In contrast to other species, it has paler green leaves with a very short, narrowly winged stalk clearly distinguished from the blade. The flowers of the lemon have a strong, pleasant scent and are produced throughout the whole year, so that various stages of flowering and fruiting are visible on a plant at the same time.

## Lime

*Citrus aurantiifolia* (Christm.) Swingle

**Rue family** Rutaceae

The lime bears a certain similarity to the lemon. It originates in S.E. Asia and is cultivated mainly in tropical regions, especially in Central America. It produces yellow fruits which are similar to lemons but considerably smaller. The fruit contains less citric acid, but there are other aromatic substances present. In appearance, the plant resembles the lemon tree but is smaller. It is usually grown in gardens attached to houses, but frequently escapes and becomes naturalised. As a rule the fruits are consumed in the producer country and are rarely exported. Lime juice was drunk to prevent scurvy on long sea voyages and like the other citrus juices is a popular flavouring and cocktail mix.

## Citron

*Citrus medica* L.

**Rue family** Rutaceae

The citron is native in the subtropical mountain districts of India, and produces fruits with a very thick, wrinkled or warty rind which is rich in ethereal oils. The mesocarp of the unripe fruit is first sliced and fermented in brine. It is then placed in syrup and candied. This is the form in which it is usually found in commerce. The citron is a small tree, scarcely exceeding 3m in height, with reddish stems bearing short, stout spines, and the leaf-stalks are usually unwinged. The petals are white on the inner surface, and tinged pink on the back. The citron is also cultivated in the southern mediterranean region.

## Pomelo, Shaddock

*Citrus grandis* (L.) Osb.
Syn. *C. maxima* (Burm.) Merr.

**Rue family** Rutaceae

The pomelo occurs both wild and in cultivation in its native region of south-eastern Asia, and is also grown in other parts of the tropics. It is remarkable among *Citrus* species for the size of its fruits which can be up to 25cm in diameter and can weigh as much as 5–6kg. Underneath the very thick, soft yellow mesocarp is the flesh, which falls apart easily but has a relatively bitter taste, so it is usually eaten only as a dessert fruit. In spite of this it is often grown for its strikingly large fruits. It forms a tree 5–10m high, but is rather tender and has stems which are hairy when young. Spines are sometimes produced or may be altogether absent. The large oblong-ovate leaves are usually rounded at the apex and have broadly winged stalks.

## Grapefruit 258

*Citrus paradisi* Macf.
Syn. *C. decumana* L. var. *racemosa* Roem.

**Rue family** Rutaceae

The grapefruit arose in the tropics as a cultivated plant but its precise origin is not known for certain. It seems to have appeared more than 200 years ago in the West Indies, though it is more suited to a temperate climate. Deliberate cultivation of the plant as a fruit tree began about 1880 in Florida. It is now grown in increasing quantities in Israel and South Africa too. The flesh of the fruit is characterised by a pleasant, slightly bitter taste, due to the glycoside naringin. The tree grows up to 10m in height, and, like the pomelo, its leaves have broadly winged stalks. Its flowers are typical of all citrus fruits, having numerous stamens joined into a tube round the style with its capitate stigma. Because its true origin is unknown, there is some doubt as to whether it is a true species, and, in fact, it may have been derived from the pomelo.

## Sour Orange, Seville Orange — 259

*Citrus aurantium* L.

**Rue family** Rutaceae

The sour orange is native in northern India. It is the hardiest of all the common citrus fruits and is therefore used as a rootstock for grafting. Botanically it is a small, low-growing tree with rather slender thorns on its branches. The leaf-stalks are clearly winged, and the blades are oblong-ovate in shape. The fruits are globular, and deep orange in colour when fully ripe. The peel is rough and thick, and the pulp sour and bitter, but when cut up or grated it provides the basis for orange marmalade. The ethereal oil obtained from the flowers and fruits, known as bigarade oil, is of great economic importance, and has many uses in the cosmetic industry. The variety *bergamia* is the source of bergamot oil which is used in the manufacture of liqueurs and for medicinal purposes, and the plant itself is called the bergamot orange. Like the citron, the peel of the sour orange is often candied. Present-day centres of cultivation include Spain, Sicily, Morocco, India, Paraguay, South Africa and the West Indies.

**Production of Citrus Fruits** (in thousand tons)

| *Country* | *Oranges* | | | *Mandarins/ Clementines* | | | *Lemons* | | |
|---|---|---|---|---|---|---|---|---|---|
| | 1975 | 1978 | 1980 | 1975 | 1978 | 1980 | 1975 | 1978 | 1980 |
| U.S.A. | 9294 | 8643 | 10740 | 619 | 613 | 756 | 1053 | 916 | 756 |
| Brazil | 6333 | 7818 | 8948 | 310 | 330 | 102 | 70 | 90 | 469 |
| Mexico | 2322 | 2400 | 1630 | 156 | 71 | 504 | 623 | 440 | 180 |
| Spain | 2016 | 1648 | 1741 | 652 | 852 | 314 | 254 | 239 | 968 |
| Italy | 1582 | 1400 | 1830 | 350 | 305 | 802 | 880 | 652 | 355 |
| Israel | 1052 | 919 | 850 | — | 85 | 63 | 38 | 29 | 84 |
| India | 940 | 1037 | 1150 | — | — | 405 | 450 | 453 | — |
| Egypt | 856 | 700 | 1092 | 97 | 80 | 71 | 59 | 47 | 100 |
| China | 818 | 838 | 894 | 228 | 264 | 84 | 59 | 65 | 256 |
| Argentina | 729 | 650 | 762 | 230 | 240 | 370 | 339 | 300 | 222 |
| South Africa | 600 | 591 | 550 | — | — | 30 | 25 | 35 | — |
| Turkey | 551 | 660 | 692 | 105 | 135 | 200 | 290 | 330 | 140 |
| Greece | 538 | 600 | 596 | 40 | 40 | 177 | 191 | 169 | 36 |
| Morocco | 477 | 643 | 720 | 106 | 142 | 3 | 10 | 5 | 267 |
| Japan | 387 | 353 | 401 | 3823 | 3097 | — | — | — | 2967 |
| Algeria | 337 | 285 | 305 | 149 | 150 | 8 | 10 | 9 | 149 |
| Pakistan | — | 530 | 515 | — | 200 | 30 | — | 30 | 200 |
| Ecuador | 275 | 520 | 505 | 28 | 56 | 16 | 32 | 16 | 29 |
| Other countries | 3595 | 3875 | 4877 | 316 | 376 | 865 | 710 | 820 | 427 |
| Total | 32702 | 34110 | 38798 | 7209 | 7036 | 4870 | 5093 | 4645 | 7605 |

## Pomegranate

260, 261

*Punica granatum* L.

**Pomegranate family** Punicaceae

Because it has attractive flowers as well as edible fruits, the pomegranate is both an ornamental and an economic plant. Its natural distribution extends from Persia to the Hindu Kush, though nowadays it is cultivated in all the tropical and subtropical countries of the world. Pliny called the plant 'malum punicum' (Punic apple), and the Egyptians and Greeks regarded it as sacred. The Greeks also considered it as a symbol of fertility because of the large number of seeds in each fruit. The words 'grenade' and 'garnet', also the name of the Spanish town of Granada are all derived from the name of this plant.

The cultivated pomegranate grows up to 6m high, forming a tree with hard wood and splitting bark (Photo 260), but in the wild it is often only a low, spiny shrub. Its light brown branches have a number of short shoots which produce smooth, evergreen, lanceolate leaves singly or in clusters. The brilliant red flowers (Photo 261) which arise in the leaf-axils are single in the wild form and double in cultivated varieties. They have a fleshy, five- to seven-lobed calyx and numerous stamens. The inferior ovary resembles that of the navel orange in having two or three whorls of carpels placed one above another. This arrangement is clearly visible in the ripe fruit. The fruit is a globular berry, brownish orange in colour and tinged red to a varying degree. It has a leathery skin and is crowned by the persistent calyx. The abundant seeds formed within the carpels are surrounded by a fleshy, jelly-like pulp which is the edible portion of the fruit. Recently, the juice has been extracted to form the cordial 'grenadine', used in making cocktails, and the pulp has been processed to produce jam and syrup. The bark of the trunk and the roots are rich in tannin and are used for medicinal purposes. The tree will thrive in a subtropical climate if the ground is irrigated and will tolerate salt to a certain extent. In addition to *P. granatum* there is, on the island of Socotra, a species called *P. protopunica* which may represent the form from which the cultivated plant has been derived.

## Date Palm

262

*Phoenix dactylifera* L.

**Palm family** Palmae

The date palm is not a typical tropical plant but is characteristic of the marginal regions of the tropics where it is one of the oldest cultivated plants. This is confirmed by pictorial representations nearly 3000 years old. At the same time they provide information about the origin of the date palm, which probably originated in the hot, dry regions round the Persian Gulf. It is possible that it is derived from *P. silvestris*, a wild species native in that area, though other interpretations are equally valid. The palm has a slender trunk, 20–30m high, covered with the remains of leaf-bases and crowned by a tuft of 30–40 pinnate leaves up to 4m in length. The larger leaflets are divided at the tips, and the lower ones are often reduced to thorns. This species resembles the closely related Canary date palm in having male and female flowers on separate trees. Its paniculate inflorescences arise from the axils of the leaves and before opening are enclosed by two bracts. Female trees have several hundred rather insignificant flowers in each panicle. Each flower has perianth-segments in two whorls of three, and the three carpels are not united. In contrast, male trees produce panicles with twice as many flowers. Only one of the three carpels develops into a fruit, the date, which contains a single seed, the 'stone', furrowed lengthwise. Botanically considered to be a one-seeded berry, the date is golden yellow or brownish red in colour and the bunches of fruit glisten in the strong sunlight.

The date palm grows best in hot places with an average summer temperature of 30°C. According to the words of an Arab poet the tree has 'its foot in the water and its head in the fire of heaven'. As it has been in cultivation since ancient times, there are numerous forms. In simple terms, these can be divided into 'soft dates', which have a high sugar content and are eaten raw, and 'dry dates', which are rich in starch and form the basic food of desert dwellers. This is easily understandable when one realises that there are up to 40 different cultivated varieties in the oases, ripening at different times. In cultivated areas, one male tree is left standing for every 30–50 female plants. Since ancient times, pollination of the female flowers has been carried out by shaking male inflorescences into the female tree as soon as the flowers have opened, or, since the pollen remains viable for several months, by hanging male inflorescences amongst the female ones. Cotton cloths, saturated with pollen, have also been used. After a ripening period of five or six months, the tender 'soft dates' are picked individually, while the 'dry dates' are allowed to reach complete maturity. Yields vary between 10–50kg (in exceptional cases 150kg) per tree annually. The date palm is usually propagated by cutting offshoots away from mother plants which bear good crops of fruit. Young plants begin to fruit when they are eight years old, sometimes even earlier, and produce their highest yields from 30 years onwards. These continue until they are about 80, although the trees themselves can live to be 200 years old. Since the end of the 19th century, cultivation of date palms has no longer been restricted to their native region from India across Persia to north-western Africa. They have been grown successfully also in California, Texas, Mexico, Brazil, Argentina, South Africa and Australia. Apart from great heat, date palms require plenty of underground water during fruiting time. Light showers of rain during the flowering period can result in a reduced yield of fruit. The plant is tolerant of alkaline soils, a fact which is confirmed by the large number of cultivated palms in ancient settlements. Apart from the fruits, the young leaves are eaten as a vegetable and the trunks used as timber for building. The trunk of *P. silvestris* can be tapped for its juice which is fermented into palm wine.

**Production of dates**
(in thousand tons)

| *Country* | *1975* | *1978* | *1980* |
|---|---|---|---|
| Iraq | 496 | 581 | 395 |
| Egypt | 415 | 460 | 418 |
| Iran | 290 | 300 | 300 |
| Saudi-Arabia | 262 | 300 | 422 |
| Algeria | 182 | 196 | 180 |
| Pakistan | 175 | 210 | 205 |
| Sudan | 102 | 109 | — |
| Other countries | 477 | 508 | 820 |
| Total | 2399 | 2664 | 2740 |

# Industrial Plants

## Cotton

263, 264, 265

*Gossypium herbaceum* L., *G. hirsutum* L., *G. vitifolium* Lam. and *G. arboreum* L.

**Mallow family** Malvaceae

Cotton is not a purely tropical economic plant. On the contrary its cultivation extends from latitude 47° north to 28° south of the equator. It is not only a soft fibre plant but is a source of oil. It was being grown in the Indus valley before 3000 BC and in Peru before 2500 BC.

Nowadays the genus *Gossypium* is distributed throughout all tropical regions of the world, and wild forms are known from South Africa and Indonesia on the one hand and the western slopes of the northern Andes on the other. A common ancestral form cannot now be traced. Consequently, species of cotton grown nowadays have various centres of origin. The word 'cotton' is derived from the Arabic 'kutun'.

Cotton as a commercial product is obtained from the unicellular, flattened and twisted hairs which form a covering for the seeds. Most kinds of cultivated cotton plants are herbaceous perennials, which are maintained as annuals in cultivation. However, truly annual races are also used, most of them derived from *G. herbaceum*.

From an ecological viewpoint it is worth mentioning that the genus *Gossypium* comprises species which occur in both dry and wet localities. Descendents of the first group are in the majority, and these are cultivated in areas with definitely dry seasons. Those of the second group thrive in humid forest regions and in river basins with abundant rainfall, e.g. the lower Amazon and Orinoco. All races are susceptible to frost. In areas with a seasonal climate, they require sufficient amounts of water during their early development, but after that, dry weather is necessary. As the cotton plant has a tap-root 2–3m in length, it needs a deep, well aerated soil, not too rich in mineral salts. On the coasts of Yucatan there is a creeping form belonging to the dune vegetation which is very resistant to salt spray.

Typically, the cotton plant is an erect herbaceous perennial with alternate leaves, usually five-lobed but sometimes three-lobed or heart-shaped and undivided. From the axils of the upper leaves yellow, white (Photo 263) or purplish flowers appear which are similar in structure and outward appearance to *Lavatera*, which belongs to the same family. Fertilisation occurs mainly after self-pollination, and only occasionally through visiting insects. The capsule takes about four weeks to ripen, and during this period hairs develop on the surface of the pear-shaped seeds. Two kinds of hairs are recognised, the longer, up to 40 or even 65mm, are called 'lint' and the shorter are known as 'fuzz'. The length of lint is further distinguished by referring to that over 28mm as 'long staple' and anything under 25mm as 'short staple'. At maturity the capsules split from the top along the edges of the carpels, exposing the dark brown seeds and their hairs which are usually white (Photo 264).

The species most often grown are two old-world species (*G. arboreum* and *G. herbaceum*) and two new-world species (*G. hirsutum* and *G. vitifolium*). In addition there is a wide range of hybrids and selected forms.

Harvesting (Photo 265) of high quality cotton is carried out by hand in Egypt and the Sudan. In the USA and the Soviet Union machines are used which reduce the quality as the capsules do not all ripen simultaneously and leaves may also be gathered. Recently, new types of machines have been developed in the USA, which allow for variation in harvesting. Yields vary between 0.1–0.7 tons per hectare of lint and 0.2–1.4 tons per hectare of seeds. Details are given in the table below.

The fibres consist of up to 90% cellulose and are valued according to their length or staple. Long staple varieties are usually grown in Egypt and Peru, medium staple in the USA, and short staple in Asia. Medium-stapled cotton accounts for about 75% of the total production, about 15% is long-stapled, and 10% short-stapled. After harvesting, the cotton passes through machines which remove first the lint, and then the fuzz from the seeds. The long fibres are spun and the fuzz used as stuffing. Because of its high cellulose content, the short fibres can be used in the manufacture of paper, cellulose and artificial silk.

Apart from the fibres, the seeds, about 3mm in length, represent an important subsidiary product in cotton cultivation. They contain 16–24% of fatty oil, 15–34%

**Production of Cotton**
(in thousand tons)

| *Country* | *1975* | *1978* | *1980* |
|---|---|---|---|
| Soviet Union | 2649 | 2640 | 3200 |
| China | 2385 | 2100 | 2707 |
| U.S.A. | 1807 | 2360 | 2422 |
| India | 1160 | 1250 | 1400 |
| Brazil | 531 | 460 | 578 |
| Pakistan | 514 | 548 | 700 |
| Turkey | 480 | 515 | 460 |
| Egypt | 382 | 435 | 530 |
| Sudan | 229 | 167 | 114 |
| Mexico | 197 | 332 | 340 |
| Argentina | 172 | 228 | 146 |
| Syria | 142 | 145 | 127 |
| Iran | 139 | 150 | 70 |
| Colombia | 139 | 82 | 101 |
| Greece | 130 | 138 | 115 |
| Nicaragua | 123 | 144 | — |
| Guatemala | 120 | 133 | 156 |
| El Salvador | 78 | 74 | 65 |
| Chad | 65 | 54 | — |
| Peru | 63 | 81 | 98 |
| Afghanistan | 53 | 54 | — |
| Nigeria | 52 | 37 | — |
| Israel | 49 | 77 | 78 |
| Tanzania | 45 | 56 | 51 |
| Mali | 39 | 43 | 48 |
| South Africa | 35 | 47 | 52 |
| Paraguay | 32 | 81 | 75 |
| Other countries | 552 | 520 | 758 |
| Total | 12362 | 12951 | 14391 |

protein and 21–33% carbohydrates. A brown oil is extracted from the thick-coated seeds which is of great importance in the manufacture of margarine.

This is the oil that is obtained by the first pressing. Subsequent pressing produces an industrial oil that is important for the manufacture of soap, cosmetics and candles or is used as lubricating oil. It contains the brown, very poisonous substance 'gossypol' which is rendered harmless by exposure to heat or even to the air. It is also present in the residual seed cake, which contains 6–16% oil and 23–44% protein. This is a valuable food for cattle and other domesticated ruminants since the minute quantities of gossypol remaining do them no harm. Processes are being developed whereby the poisonous substance can be entirely removed and the residue turned into cotton-seed flour and used for human nourishment. At the present time about 20 million tons of cotton seeds are processed annually.

## Kapok, Silk-cotton Tree — 266

*Ceiba pentandra* (L.) Gaertner

**Cotton-tree family** Bombacaceae

The kapok tree is native in South and Central America and occurs not only in the rain-forests but also in forest regions with a seasonal climate where there are varying amounts of moisture. In the ancient Indian cultures of Central America the tree had a great mythological significance, and Nicaragua, a country of mestizos, includes the *Ceiba* in its national arms. Nowadays it is found throughout the tropical regions of the world, though systematic cultivation began only recently. When primary forest is being cleared in Central America, it is often the kapok trees that are allowed to remain for future use.

The kapok tree is a large tree, and when young, its trunk is densely covered with sharp, conical spines. Old trees often have large plank buttresses, a metre or so high, and umbrella-like crowns. The branches are frequently colonised by epiphytes (mainly bromeliads and orchids) and cacti, and provide a home for iguanas. Kapok trees bear digitate leaves, red at first on account of the formation of anthocyanin, which usually fall at the beginning of the dry season although some may remain on the branches. Then white or pink flowers with numerous stamens appear in clusters at the ends of the twigs. After pollination by bats, the five-celled ovary develops into a pointed capsule, up to 15cm long, which resembles a cocoa pod. The capsule contains up to 100 round, brownish black seeds, completely covered by soft, silky, yellowish white hairs. These fibres are outgrowths from the inner wall of the capsule – not from the outer covering of the seed as in cotton – and come away when the fruit is ripe. The fibres are 10–35mm long, and, because of a waxy coating, they are lustrous, water-repellent and unsuitable for spinning. They have a cellulose content of up to 64% and are very elastic. The fibres are therefore valuable as insulating material and for stuffing such articles as life-jackets and life-belts.

The kapok tree thrives best in regions with a seasonal climate where there is sufficient light and also a rainfall of 1500mm. It fruits from its fifth year until it is about 60, producing annually up to 20kg of pure fibre. The main producers of kapok fibre are Indonesia, with its monsoon climate, Madagascar, Cambodia, South and Central America, and total world production amounts to about 30,000 tons per year. The seeds contain 22–25% of fatty oil, which is almost identical to cotton-seed oil and is used for the same purposes. The Indian silk-cotton tree, *Bombax malabaricum*, which is native in S.E. Asia, is related to the kapok tree but its fibre is not of such high quality. In spite of this, it accounts for almost half of the total production of kapok. Another species is *B. emarginatum* which is only found in the karst country of western Cuba and is endemic to that island.

## Sisal

**267, 268, 269**

### *Agave sisalana* Perrine

**Daffodil family** Amaryllidaceae

Sisal is named after a port in the Gulf of Mexico, and it is native in Mexico and Central America. Within this area there are several hundred species of *Agave* which are difficult to distinguish. Recent cultivation of agaves started in Yucatan, but their usefulness was recognised even in ancient times. It is known from relics found in graves in Mexico that the coarse fibres of the *Agave* were being used 8000 years ago for making fishing nets and cords. Valuable pioneer work in extending the cultivation of *Agave* was rendered by the German East Africa Company. The first *Agave* plants reached Tanganyika in 1893, and cultivation there experienced a rapid rise through the German experimental stations.

Of all the species of *Agave*, sisal is the most important as a source of hard fibre. The plant has a short, thick stem bearing a rosette of stiff, bluish green, lanceolate leaves up to 2m long (Photo 268). The leaves are tipped by a spine as hard as a splinter of glass which frequently causes injuries to the harvesters. The sisal plant lives for 8–15 years and in the course of its life forms a stem 1m in length. Each year 12–15 leaves are produced which are ready for cutting after two years. In harvesting the leaves are cut off at the base by hand, but it is important that enough leaves remain on the plant so that there is no serious check to its development.

When it is 12–15 years old, the plant changes from a vegetative to a flowering phase, producing a branched inflorescence 7–8m high composed of numerous flowers. The flower resembles that of a tulip, except that the three-celled ovary is inferior. Agaves rarely produce capsules, since after successful fertilisation, the ovaries fall off prematurely. Instead, bulbils develop on the flowering stem, and these, together with the suckers which arise from the base of the parent plant, can be used for vegetative propagation in the plantations (Photo 267). The bulbils can be shaken off the stem or the stem itself can be cut in order to reach them. In this case, a juice containing starch and sugar oozes from the cut and is fermented to form a highly intoxicating liquor known in Mexico as 'pulque'. After flowering the plant dies.

Since agaves are tropical plants they need a hot climate in order to grow well. The temperature may exceed 30°C but must not fall below 15°C. In large areas of Central America they are grown on relatively poor soils in districts characterised by a definite hot, dry season.

The coarse fibres, 1–2m long, are obtained from the leaves through a process of crushing, while the non-fibrous tissue is washed away. After further washing and drying the glossy, yellowish white fibres are rendered pliable by beating and brushing. They are the basic material used in the manufacture of twine, ropes, ships' cables, also nets, hammocks, upholstery and carpets (Photo 269). Fibre production varies between 0.2–1.7 tons per hectare annually, but has declined now that synthetic fibres are available.

In addition to sisal, the Mexican *A. fourcroydes* or henequen is an important source of fibre, but as its leaves are edged with spines harvesting presents problems. This species lives to be 30 years old, forming a stem 2m high and 25cm in diameter which is crowned by a rosette of leaves. Its centres of cultivation are in Mexico and Cuba where the coarser leaf-fibres are made mainly into twine. Another important fibre-producing plant is cantala, *A. cantala*, which is grown mostly in the Philippines and Java. In El Salvador *A. letonae*, the Salvador henequen, is grown for its finer fibre, and in the Andes the closely related genus *Furcraea* is used to obtain 'pita' fibre. When plants of this genus are in flower they have have a very tall, branched infloresence, the pendulous lateral branches of which bear thousands of flowers.

**Production of sisal**
(in thousand tons)

| *Country* | *1969–71* | *1975* | *1980* |
|---|---|---|---|
| Brazil | 314 | 202 | 254 |
| Tanzania | 128 | 125 | 115 |
| Kenya | 48 | 31 | 48 |
| Angola | 40 | 20 | 20 |
| Madagascar | 21 | 17 | 19 |
| Mozambique | 15 | 18 | 18 |
| Venezuela | 14 | 20 | 14 |
| Haiti | 12 | 13 | 16 |
| Other countries | 18 | 20 | 12 |
| Total | 610 | 466 | 516 |

## White Jute
*Corchorus capsularis* L.

## Tossa Jute, Upland Jute
*Corchorus olitorius* L.

**Lime family** Tiliaceae

The jute plant was originally known in Asia as a vegetable as its shoots and leaves are very tasty. This applies only to white jute. Its use as a fibre plant was very limited, and at first it was cultivated only in the gardens of houses in south-east Asia and India. When in 1828 some bales of this soft fibre were shipped for the first time from Calcutta to England, appropriate spinning and processing techniques were developed, and jute began to grow in importance as a source of fibre.

Wild forms of these two plants have not definitely been found, but they are probably native in India or China, and closely related forms undoubtedly occur there. However, it should be remembered that the plant has been grown since immemorial times as a culinary vegetable, chiefly in gardens of houses in Bengal. Since the jute-plant, especially in its young state, requires high humidity and a temperature about 30°C, it grows best in the river-basins of the Ganges and the Brahmaputra in India. There, the soil, a sandy loam, is flooded annually with organic nutrients, but is always well-

Pieces of stem of white jute, *Corchorus capsularis* (left) and ramie, *Boehmeria nivea* (right).

drained. Sowing is usually carried out by hand and individual plants are transplanted like rice.

The jute plant reaches a height of 4m and is somewhat like a hollyhock. Its stems are slender, about the thickness of one's finger, and bear alternate, short-stalked, lanceolate leaves whose base is drawn out into two long teeth (see illustration above). The yellowish white, five-petalled flowers, small and inconspicuous, appear singly or in small clusters from the nodes of the stem. After self-fertilisation the superior ovary develops into a capsule. White jute has small, almost globular, slightly wrinkled fruits, while in tossa or upland jute they are cylindrical and 6–8cm long. The first species is decidedly tropical in its requirements, while the second will thrive also in the higher, cooler foothills of the Himalaya. White jute is so called because of its very pale coloured fibre. Tossa jute, on the other hand, has yellowish or reddish fibre. The first species may be recognised by its edible leaves, shiny on the upper side. The other species has dull leaves which are unpalatable on account of the bitter substances they contain.

Cultivation, harvesting and processing of jute is very laborious. The tough, rather bushy plants are cut through with a sickle at the base of the stem when the fruits begin to ripen. At this time there are often monsoon floods in eastern India and the harvesters may even have to dive under water to cut and gather the plants. The stems are then tied into bundles and submerged in slow running water where micro-organisms break down the soft tissue. After eight to ten days, the stems are placed in still water and the process of separating the fibres completed by hand. Then the fibre cells, which are associated with the vascular bundles in the stem and are only 2mm long, are combed to produce fine fibres 1–3m in length and suitable for spinning.

Tossa jute, with its soft, yellowish or reddish fibres, produces a finer, more shiny fabric, but two-thirds of Indian jute comes from white jute plants. Unfortunately the fibres cannot be bleached easily, but they can be coloured without difficulty, so that cheap and pretty items of clothing can be manufactured from them. The fibre is also used for making sacks, cloth backing for linoleum, carpets, screens, tents, coverings for cables, and twines.

## Ramie, Rhea, China Grass

*Boehmeria nivea* (L.) Gaud.

**Nettle family** Urticaceae

Ramie is the most important of all fibre plants. The fibres are considered more valuable than those of any other plant because of their high tensile strength, the yarn spun from them being eight times stronger than cotton. The plant is native in eastern Asia and was cultivated there before cotton. It was first grown in Europe, North Africa and North America in the 19th century, and more recently also in Brazil and the Philippines.

The plant is a perennial, related to the nettle, with an underground rootstock which sends up stems over 2m high with only a few branches. The long-stalked, heart-shaped, sharply toothed leaves are arranged alternately on the stems. In white ramie (var. *nivea*) the leaves are white-felted on the underside, while in green ramie (var. *tenacissima*) they are green. Panicles of inconspicuous, unisexual flowers arise from the axils of the leaves (see illustration on p. 240), and are pollinated by the wind. The plant requires high rainfall and a deep soil, rich in nutrients. It can then be cut up to six times a year. The fibres are cells situated towards the outer surface of the stem for strengthening purposes. They are removed fresh or in a dry state by hand or by machines, and this product arrives at the factory as China grass. The hard fibres, 12–25cm long, have a gummy covering, rather like pectin, which cannot be removed by the usual steeping process involving bacteria. Removing this gum from the fibrous bundles is very costly and is achieved by treating them with soap solution, lime or chemicals. The fibres, containing 69% cellulose and 13% hemicellulose, are highly resistant to water and are cool to the touch. Consequently they are in great demand in the tropics for the manufacture of underwear and bed-linen. They are also used in making plush, straining cloths, fire-hoses, and special fabrics such as that used for parachutes. The main producers are China, Brazil, Japan, Philippines, Taiwan, Cambodia and Korea. In contrast to the Chinese variety of ramie (var. *nivea*), the Malayan variety (var. *tenacissima*) is unquestionably a warmth-loving plant, and will not tolerate any cold. It is sometimes treated as a separate species and called *B. utilis*.

## Papyrus 270

*Cyperus papyrus* L.

**Sedge family** Cyperaceae

Papyrus was an important economic plant in ancient times, but nowadays it is much more often grown as an ornamental. The generic name of the plant is derived from the Greek word ‘kypeiros’. The genus *Cyperus* comprises some 600 species, distributed almost entirely in the tropical and subtropical regions of the world. Without any doubt, the best known species is the papyrus plant, which originated in the tropics of central Africa. From there it was spread by man not only to Egypt, but also to Syria, Asia Minor, Sicily and Calabria. Nowadays, rivers in the African interior are bordered by kilometre-wide stands of papyrus. It is a perennial with a stout rhizome that creeps along in the mud. The bluntly triangular stems which arise from this rootstock are 4–5m high and more than 10cm across. At the top is a tuft of fine, long, pendulous leaves, and an umbel of flowers. The inflorescence has a hundred or more rays and where these meet the stem there is an involucre of eight bracts. The spikes which make up the umbel are in clusters of three to five. In ancient times the pith of the stem was made into an early form of paper (a word which comes from the Latin papyrus). For this purpose, the pith was cut in strips and laid side by side with the edges overlapping. The next layer was laid crosswise. This 'paper' could only take writing on one side as it was not sized. The Greeks called the material 'byblos', a word that has been preserved as 'bible' (book). In Egypt the pith of the plants was being used in the manufacture of papyrus even before 2400 BC. The starchy rhizomes of the plant used to be eaten, and the outer covering of its stem used in making cords, mats and baskets.

## Red Mangrove — 271

*Rhizophora mangle* L.

**Mangrove family** Rhizophoraceae

*Rhizophora* is a genus of the Old and New World tropics. In tidal regions along the tropical coasts, especially in the neighbourhood of river estuaries, mangroves form thickets and even forests of varying height and structure. Stands in the Indomalayan archipelago can reach a height of more than 30m. There are more species of mangrove in Asia and Africa than there are in America, and they differ in their aerial roots. All mangroves are more or less resistant to the effect of sea salt, and are biologically among the most intriguing forms of plant life, showing a whole range of highly interesting adaptations to their environment. Adaptation of the various species to their particular locality can be seen in the formation of stilt-roots and aerial roots. In *Rhizophora*, stilt-roots are especially well formed and allow the plant to extend into the sea further than any other plant. Also characteristic of this genus is the fact that the seeds begin to germinate while the leathery fruits are still attached to the mother-plant. The young plants, conical in shape and resembling a massive arrow, grow up to 30cm long (in the east asiatic *R. mucronata* 50–60cm), then fall to the ground, boring into the mud where they land. Apart from mangroves with stilt-roots, which form impenetrable thickets round tropical islands, there are others with aerial roots which arise from roots creeping underneath the surface of the mud, and poke up at low tide like beds of asparagus.

Stems and roots of all mangroves contain 10–40% of tannins in their bark and these substances, used in processing leather, are obtained from wild plants. The timber is used for building purposes and is also a source of charcoal.

## Pará Rubber — 272, 273

*Hevea brasiliensis* (Willd. ex Adr. Juss.) Muell. Arg.

**Spurge family** Euphorbiaceae

So far, more than 1000 species of plants from 80 genera have been discovered in the various tropical regions of the world which produce a milky juice capable of being turned into rubber. But the one which is of world-wide importance is *H. brasiliensis* which occurs wild south of the river Amazon. A closely related species from north of the Amazon is known as *H. benthamiana*, after the botanist Bentham. There are also other cultivated *Hevea* species, and the natural distribution of these plants is shown in the figure below. To a large extent, the plants in cultivation nowadays are derived from *H. brasiliensis*.

The name of the botanical genus *Hevea* is derived from the word 'heve', the name given to a rubber-producing tree by the native inhabitants on the upper Orinoco in the vicinity of Esmeralda. Other Indian tribes in the rain-forests of the interior of Brazil called the rubber-plants 'caá-huchu', meaning 'weeping trees'. The first description of rubber goes back to 1521 when the Spaniard Pedro Martir de Anghiera noticed the Indians on Haiti playing with balls of rubber which had been made from the secretion of a tree called 'ule'. Humboldt and Bonpland made the first scientific collections of rubber-trees on their famous journey of exploration in 1799 and 1800 on the upper Orinoco. Careful tests have shown that the material collected should be called *H. pauciflora* a closely related species from Brazil.

For a long time the production of rubber was restricted to countries such as Brazil, Peru, Venezuela and Colombia where rubber-trees are native. It was not until 1877 that an Englishman, H.A. Wickham, smuggled 70,000 seeds of *H. brasiliensis* from the state of Pará to England on board the steamer 'Amazonas'. The shipment was received by the famous botanist and director of Kew Gardens, Sir Joseph Hooker. From these seeds he raised young plants which were shipped to Singapore and formed the basis of

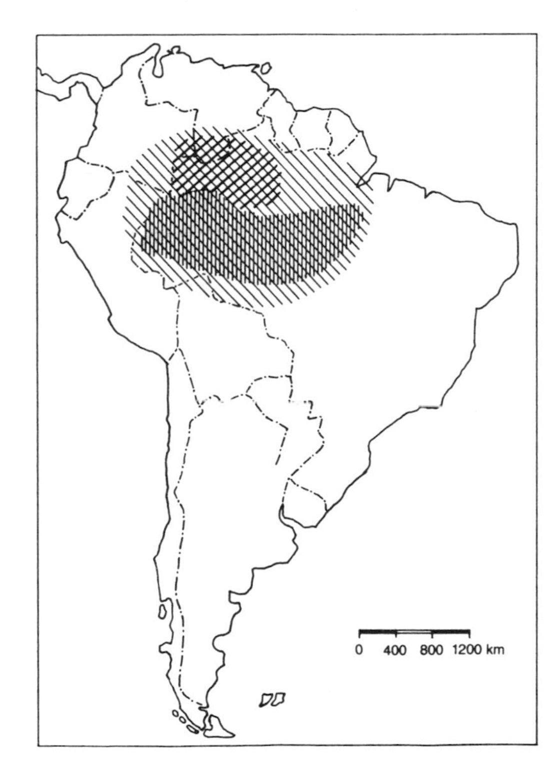


Natural distribution of *Hevea* species (according to H. Brücher, *Tropical Economic Plants*, Springer-Verlag, 1977).

 *H. benthamiana*

 *H. brasiliensis*

 Other *Hevea* species occurring wild.

Twig of the rubber-tree (*Hevea brasiliensis*) with inflorescences and fruit.

rubber-growing in Indonesia. Production from this region has now far surpassed that of the original countries.

*H. brasiliensis* is a tree up to 30m high with a smooth, light grey bark, and long-stalked leaves made up of three leaflets (Photo 272). The leaves are shed during the dry season. The unisexual flowers are arranged in panicles (see fig. below) and are wind-pollinated. After fertilisation, the three-celled ovary develops into a capsule which, when ripe, explodes noisily into six pieces, flinging the seeds a distance of about 13m. The botanist Baldwin, who spent years researching into the classification of rubber-producing trees in the Amazon region, reported that he was able to distinguish individual species merely by the sound of the exploding capsules! In order to thrive, *H. brasiliensis* requires an average temperature of 27°C, not sinking below 18°C, about 2000mm rainfall per year, and a moist, well-aerated soil.

In order to obtain the latex, half-spiral or herring-bone incisions are made in the bark of the trunk, taking care not to cut so deeply as to injure the latex cells (Photo 273). A cup is wired to the tree and after a few hours it is found to be full of milky sap which has oozed from the lower end of the cuts. New cuts are then made below the older ones to ensure a continuous flow of latex.

Tapping begins when the tree is five to seven years old and can be continued for 30 years or more. Cuts are made either daily or every two or three days, and the quantity of latex obtained depends very much on the skill of the tapper, who can deal with up to 700 trees in a day. The latex contains up to 33% rubber, 2% resins and 1.8% protein and residual water. Immediately after the collecting cups have been emptied, the liquid is taken to the factory. There, ammonia is added to prevent coagulation and it is in this form that it is transported by tanker lorries or ships to the consumer countries. It can also be processed to form raw rubber by the addition of acetic and formic acid which causes coagulation of the droplets. This takes place nowadays in long tanks, which are divided up by a number of perforated partitions set close together. By this means, a long band of raw rubber is obtained, which is then washed, rolled, cut into pieces, and subjected to smoke for several days at a temperature of 45°–60°C. The rubber obtained in this way from *H. brasiliensis* is exported from Brazilian ports as Pará rubber.

To manufacture rubber the raw product must be vulcanised with sulphur. The addition of 4–5% sulphur results in an elastic, soft rubber, while adding 25–30% sulphur produces a hard rubber which does not conduct electricity. The process of vulcanisation was developed in the USA in 1839 by Charles Goodyear and was the decisive step in paving the way for widespread industrial use of raw rubber. But highly civilised Indians such as the Mayas and the Aztecs had already learned how to make protective rain-capes, waterproof shoes and rubber balls from latex.

In addition to Pará rubber, there is Ceara rubber, which is obtained from the latex of *Manihot glaziovii*. This tree is also a member of the Euphorbiaceae and is related to the starch plant manioc or cassava (see p. 181). Other sources of rubber include *Ficus elastica*, which produces Assam rubber, and various woody climbers or lianes. However, the rubber from these plants is not so valuable as that obtained from the Pará rubber tree.

**Production of rubber**
(in thousand tons)

| *Country* | *1975* | *1978* | *1980* |
|---|---|---|---|
| Malaysia | 1478 | 1599 | 1600 |
| Indonesia | 823 | 870 | 919 |
| Thailand | 349 | 453 | 510 |
| Sri Lanka | 149 | 154 | 155 |
| India | 136 | 158 | 145 |
| Nigeria | 95 | 90 | 60 |
| Liberia | 81 | 85 | 75 |
| Philippines | 35 | 63 | 55 |
| Zaire | 29 | 27 | 20 |
| Vietnam | 20 | 45 | 57 |
| Brazil | 19 | 30 | 25 |
| Other countries | 96 | 124 | 190 |
| Total | 3310 | 3698 | 3811 |

## Teak

**274**

*Tectona grandis* L.f.

**Verbena family** Verbenaceae

Among the numerous timber-producing trees of the tropics, the teak tree is one of the best known. It belongs to a genus comprising four species, all of which are distributed in S.E. Asia. The teak tree is found wild in this area although its distribution nowadays extends westwards to India and eastwards to Java, where it is grown in extensive forests. It is also cultivated in the tropical regions of the New World. It is a large, deciduous tree, 30–40m in height, which grows best in the drier areas, even on stony soils, and propagates itself by suckers. The trunk has a grey, deeply fissured bark. The leaves, arranged in pairs, are elliptic, pointed, and reach a length of 30cm. A red dye can be obtained from them. The open, paniculate inflorescences are composed of flowers with a five-lobed corolla, and a calyx which enlarges at maturity, forming a bell-shaped structure round the four-celled drupe. Since the wood of the tree contains silicic acid and oil it hardly ever warps. It is light brown in colour with a darker heartwood. The timber is hard but easy to split and has a strong and penetrating odour. It is excellent for boat-building, and is employed for house-building and railway carriages, and in the manufacture of furniture and flooring. It is also used for making household utensils of every kind. Teak is a wood of special value since it is not subject to attack by insects or fungi.

# Some recommended botanic gardens and parks in the tropics

## America

| | |
|---|---|
| Barbados: | Andromeda Botanic Gardens, St Joseph |
| Brazil: | Jardim Botânico do Rio de Janeiro, Rua Jardim Botânico, Rio de Janeiro |
| | Horto Botânico, Quinta de Boa Vista, Rio de Janeiro |
| Dominica: | The Royal Botanical Gardens, Roseau |
| Guatemala: | Jardín Botánico, Avenida de la Reforma Cuidad de Guatemala |
| Guadeloupe: | Jardin Botanique de Service de l'Agriculture Point-á-Pitre |
| Jamaica: | Royal Botanic Gardens (Hope), Mona, Kingston |
| | The Tropic Gardens of Montego Bay, Montego Bay |
| | Cinchona Botanic Gardens, Cinchona District |
| Colombia: | Finca de Orquídeas Ospina, Medellín |
| Mexico: | Jardín Botánico de la Universidad Nacional Autonoma de México, Viveros de Coyoacán, México, DF |
| Martinique: | Jardin Botanique, Saint Pierre |
| Puerto Rico: | Botanical Gardens of the University of Puerto Rico, Rio Piedras (near San Juan) |
| | Plant Collection, Federal Experimental Station, Mayagüez |
| | Palmas Botanical Gardens, 130 Candelero Abajo |
| St Vincent: | Botanical Gardens, Kingstown |
| Trinidad: | Royal Botanic Garden, Port of Spain |
| USA: | Foster Botanical Garden, Honolulu, Hawaii |
| | Lyon Botanical Garden, Honolulu, Hawaii. |
| | Fairchild Tropical Gardens, Miami, Florida |
| | Balboa Park, San Diego, California |

## Africa

| | |
|---|---|
| Ghana: | Botanic Gardens, Aburi |
| Congo (Zaire): | Jardin Botanique Eala, near Coquilhatville |
| Madagascar: | Jardin Botanique Tsimbazaza, Tananarive |
| Mauritius: | Royal Botanical Gardens, Pamplemousses |
| Nigeria: | Botanical Gardens of the University College, Ibadan |
| South Africa: | National Park and the Division of Botany, Pretoria |
| | National Botanic Gardens of South Africa, Kirstenbosch, Newlands, Cape Province (12km from Cape Town) |
| Tenerife: | Jardín Botánico de Orotava, Orotava |

## Asia

| | |
|---|---|
| Ceylon (Sri Lanka): | Botanic Garden, Heneratgoda, Colombo |
| | Royal Botanic Gardens of Peradeniya, Kandy |
| India: | National Botanic Garden, Lucknow |
| | Botanic Gardens of Calcutta, Calcutta |
| | Botanic Garden, Madras |
| Indonesia: | Indonesia Botanical Garden, Bogor, Java |
| | Indonesia Botanical Garden Tjibodas, Bogor, Java (Tropical Rock Garden) |
| Malaysia: | Public Gardens, Kuala Lumpur |
| | Waterfall Gardens, George Town, Penang |
| | Botanic Garden, George Town, Penang |
| Philippines: | Botanic Garden Rizal, Manila |
| | Makiling Botanic Gardens, University of the Philippines, Los Banos, Laguna |
| Singapore: | Public Parks of Faber Hills |
| | Botanic Gardens |
| Taiwan: | Botanical Garden, Taipeh |
| | Kenting Tropical Forest Botanical Garden, Heng-Chun, Pingtung |
| Thailand: | Phu-Khae Botanic Garden, Saraburi (near Bangkok) |

## Australia and Oceania

| | |
|---|---|
| Australia: | Botanic Gardens and National Herbarium, Melbourne |
| | Botanic Garden of the Northern Territory, Darwin |
| | Royal Botanic Gardens, Sydney, New South Wales |
| Papua-New Guinea: | Botanical Garden, Port Moresby |
| Tonga Islands: | Botanic Garden of New Caledonia |

# Glossary

**adventitious roots:** roots arising from an unusual place, e.g., the stem.
**anthocyanin:** a pigment producing red, purple or blue colouring according to the acidity or alkalinity of the sap.
**axil:** the angle formed by the upper side of a leaf and the stem.
**axillary:** in the axil.
**bipinnate:** pinnate, with the primary leaflets again pinnate.
**bract:** a much-reduced leaf, especially the small or scale-like leaves associated with a flower or flower-cluster.
**bracteole:** a small bract.
**capitate:** pin-headed, as the stigma of a Primrose.
**carpel:** the ovule-bearing structure which, either singly or together with other carpels, forms the female part of a flower.
**cauliflory:** the production of flowers on the stem or trunk.
**chlorophyll:** colouring matter of green parts of plants.
**chloroplasts:** minute bodies within a cell containing the green colouring matter chlorophyll.
**clone:** a group of plants that have arisen by vegetative reproduction from a single parent and which are therefore genetically identical.
**corona:** a structure occurring between the stamens and the corolla, as in Asclepiadaceae and Amaryllidaceae.
**corymb:** an inflorescence with flower-stalks of different lengths, causing the flower-cluster to be flat-topped.
**cotyledon:** the first leaf of the rudimentary plant formed within the seed.
**digitate:** resembling the fingers of a hand.
**dioecious:** with male and female flowers on separate plants.
**drupe:** a fleshy fruit with the seed surrounded by a stony layer.
**ecosystem:** a group of living organisms occurring together and dependent on each other.
**ellipsoid:** elliptic in outline with a three-dimensional body.
**elliptic:** in the form of an ellipse.
**entire leaves:** leaves with the edge not divided into lobes or teeth.
**epicalyx:** a whorl of sepal-like organs just below the true sepals.
**epiphyte:** a plant which grows on another plant but does not derive nourishment from it.
**epiphytic:** relating to epiphytes.
**follicle:** a dry fruit formed from a single carpel, containing more than one seed, and splitting open along one side only.
**glaucous:** with a waxy, greyish-blue bloom.
**inferior ovary:** an ovary situated below the point where the calyx, corolla and stamens arise.
**inflorescence:** the arrangement of flowers on the floral axis; a flower-cluster.
**internode:** the part of the stem between two nodes.
**involucre:** a ring of bracts beneath an inflorescence.
**lanceolate:** lance-shaped.
**liane:** a woody, tropical climber.
**ligule:** the flat membranous scale situated where the sheath and the blade of a grass-leaf meet.
**linear:** long and narrow, with parallel sides.

**loculicidal fruit:** a fruit which splits open along the midrib of the carpels.
**monocotyledonous:** having one seed-leaf.
**monoecious:** with male and female flowers on the same plant.
**mutant:** an individual differing from the original type because of a sudden hereditary change,.
**node:** the point on a stem where a leaf or leaves arise.
**oblanceolate:** lance-shaped, but attached at the narrower end.
**obovate:** ovate, but attached at the narrower end.
**ovate:** shaped like a longitudinal section of a hen's egg.
**ovoid:** egg-shaped.
**ovule:** a structure in the ovary which, after fertilisation, develops into a seed.
**palmate:** divided to the base into separate leaflets, all the leaflets arising from the apex of the leaf-stalk.
**panicle:** a much-branched inflorescence.
**paniculate:** arranged in a panicle.
**papilionaceous:** butterfly-shaped, as in flowers of the Fabaceae.
**perianth:** the outer, non-reproductive organs of a flower, often differentiated into calyx and corolla.
**perianth-segments:** the outer, non-reproductive organs of a flower, often differentiated into calyx and corolla.
**pinnate:** having separate leaflets along each side of a common stalk.
**raceme:** a spike-like inflorescence with the flowers stalked.
**radicle:** the rudimentary root formed within the seed.
**receptacle:** the apex of the stem from which all the parts of the flower arise.
**rhizome:** a root-like stem, lying horizontally on or situated under the ground, bearing buds or shoots and adventitious roots.
**sessile:** not stalked.
**spadix (spadices):** a spike of flowers on a fleshy axis.
**spathe:** a large bract enclosing an inflorescence, usually a spadix.
**spathulate:** shaped like a spatula.
**spikelet:** a unit of the inflorescence in grasses consisting of one or more flowers.
**staminal tube:** the tubular structure formed when the stamens are joined together.
**stipule:** a leafy outgrowth, often paired, arising at the base of the leaf-stalk.
**superior ovary:** an ovary situated above the point where the calyx, corolla and stamens arise.
**umbel:** a usually flat-topped inflorescence in which all the flower-stalks arise from the same point.
**umbellate:** having the inflorescence in the form of an umbel.
**whorl:** a ring of leaves round the stem or flower-parts round the floral axis.

# Index of Scientific Names

**Bold** figures refer to the colour plates.

# Index of English Common Names

**Bold** figures refer to the colour plates